S E M S M E S

Zurich Lectures in Advanced Mathematics

Edited by

Mathematics in Zurich has a long and distinguished tradition, in which the writing of lecture notes volumes and research monographs plays a prominent part. The *Zurich Lectures in Advanced Mathematics* series aims to make some of these publications better known to a wider audience. The series has three main constituents: lecture notes on advanced topics given by internationally renowned experts, graduate text books designed for the joint graduate program in Mathematics of the ETH and the University of Zurich, as well as contributions from researchers in residence at the mathematics research institute, FIM-ETH. Moderately priced, concise and lively in style, the volumes of this series will appeal to researchers and students alike, who seek an informed introduction to important areas of current research.

Previously published in this series:

Yakov B. Pesin, *Lectures on partial hyperbolicity and stable ergodicity*
Sun-Yung Alice Chang, *Non-linear Elliptic Equations in Conformal Geometry*
Sergei B. Kuksin, *Randomly forced nonlinear PDEs and statistical hydrodynamics in 2 space dimensions*
Pavel Etingof, *Calogero-Moser systems and representation theory*
Guus Balkema and Paul Embrechts, *High Risk Scenarios and Extremes – A geometric approach*
Demetrios Christodoulou, *Mathematical Problems of General Relativity I*
Camillo De Lellis, *Rectifiable Sets, Densities and Tangent Measures*
Paul Seidel, *Fukaya Categories and Picard–Lefschetz Theory*
Alexander H.W. Schmitt, *Geometric Invariant Theory and Decorated Principal Bundles*
Michael Farber, *Invitation to Topological Robotics*
Alexander Barvinok, *Integer Points in Polyhedra*
Christian Lubich, *From Quantum to Classical Molecular Dynamics: Reduced Models and Numerical Analysis*
Shmuel Onn, *Nonlinear Discrete Optimization – An Algorithmic Theory*
Kenji Nakanishi and Wilhelm Schlag, *Invariant Manifolds and Dispersive Hamiltonian Evolution Equations*
Erwan Faou, *Geometric Numerical Integration and Schrödinger Equations*
Alain-Sol Sznitman, *Topics in Occupation Times and Gaussian Free Fields*
François Labourie, *Lectures on Representations of Surface Groups*
Isabelle Gallagher, Laure Saint-Raymond and Benjamin Texier, *From Newton to Boltzmann: Hard Spheres and Short-range Potentials*

Published with the support of the Huber-Kudlich-Stiftung, Zürich

Robert J. Marsh

Lecture Notes on Cluster Algebras

European Mathematical Society

Author:

Robert J. Marsh
School of Mathematics
University of Leeds
Leeds LS2 9JT
United Kingdom

E-mail: R.J.Marsh@leeds.ac.uk

2010 Mathematics Subject Classification (Primary; secondary): 13F60; 05E40, 14M15, 17B22, 17B63, 18E30, 20F55, 51F15, 52B05, 52B11, 57Q15

Key words: associahedron, cluster algebra, cluster complex, Dynkin diagram, finite mutation type, Grassmannian, Laurent phenomenon, reflection group, periodicity, polytope, quiver mutation, root system, Somos sequence, surface

ISBN 978-3-03719-130-9

The Swiss National Library lists this publication in The Swiss Book, the Swiss national bibliography, and the detailed bibliographic data are available on the Internet at http://www.helveticat.ch.

Contact address:

European Mathematical Society Publishing House
Seminar for Applied Mathematics
ETH-Zentrum SEW A27
CH-8092 Zürich
Switzerland

Phone: +41 (0)44 632 34 36
Email: info@ems-ph.org
Homepage: www.ems-ph.org

Typeset using the author's TEX files: le-tex publishing services GmbH, Leipzig, Germany
Printing and binding: Beltz Bad Langensalza GmbH, Bad Langensalza, Germany
∞ Printed on acid free paper
9 8 7 6 5 4 3 2 1

Preface

Cluster algebras can now be regarded as a field of study in its own right, as is borne out by the allocation of a Mathematics Subject Classification number, 13F60, in the 2010 revision. But they did not exist before the current century: they were introduced in the seminal article [67] of S. Fomin and A. Zelevinsky which appeared in print in the Journal of the American Mathematical Society in 2002. So what has contributed to this phenomenal growth?

The notion of a cluster algebra captures a number of interrelated ideas in one beautiful setting. In one sense, a cluster algebra can be thought of as some kind of discrete dynamical system. It is defined using combinatorial data which is then mutated arbitrarily to produce altered copies of the original data. But it is also a Lie-theoretic object: the cluster algebras of finite type have a classification by the Dynkin diagrams, in a similar way to many other objects, including simple Lie algebras over the complex numbers and finite crystallographic reflection groups. The structure of the latter runs right through the corresponding cluster algebras.

A cluster algebra is also a representation-theoretic object: attempts to categorify cluster algebras, i.e. to model them using associated categories, have led to new developments in the representation theory of algebras. A cluster algebra is also a combinatorial object: the combinatorial fascination begins with interesting formulas for numbers of clusters and frieze patterns and goes on into combinatorial geometry, with new insights into generalized associahedra. In fact, a cluster algebra is also a geometric object, with strong connections to the the theory of Riemann surfaces.

In recent years, a number of excellent survey articles explaining how cluster algebras are representation theoretic objects have been published (more details are given in the introduction), but there is scope for a more detailed description of some of their other aspects. This book arose out of a desire to explain some of these. The large size of this growing field means that it is not possible to include everything that is being done, but the aim is to give a good idea of a number of interesting developments in the field.

It is intended that these notes should be accessible to graduate students. Where proofs are not included, detailed references are given to allow those who wish to learn more to delve more deeply into the subject. Thus the aim of these notes is to give an introduction to cluster algebras in their own right and to some of the many aspects of cluster algebras.

In Chapter 1, we introduce cluster algebras and describe the motivation for them. Chapter 2 gives key definitions via matrices and mutation, and first properties of cluster algebras, leading onto the definition via exchange patterns in Chapter 3, where a description of the relationship between the approaches via matrices and polynomials is given. As a preparation for the finite type classification theory, Chapter 4 is a short introduction to reflection groups. Chapter 5 gives a description of cluster algebras of finite type and their classification. Chapter 6 is an introduction to the generalized associahedra, which are polytopes associated to cluster algebras of finite type. Chapter 7 covers the notion of periodicity in cluster algebras, in the context of the Laurent phenomenon and integer sequences, as well as the categorical periodicity in Keller's proof of the periodicity conjecture of Zamolodchikov. Chapter 8 looks at quivers of finite mutation type, including cluster algebras arising from marked Riemann surfaces, and Chapter 9 considers the cluster algebra structure of the homogeneous coordinate ring of a Grassmannian.

These lecture notes began as a Nachdiplom (graduate) lecture course on Cluster Algebras given at the Department of Mathematics at the Eidgenössische Technische Hochschule (ETH) Zürich (the Swiss Federal Institute of Technology) in Zürich in Spring 2011, while I was a guest of the Forschungsinstitut für Mathematik (FIM, Institute for Mathematical Research). I am very grateful for the welcome I received in the Department and the Institute. I would also like to thank everyone who attended and contributed to the course and made it such an enjoyable experience for me.

I would like to thank Lisa Lamberti, whose typed up version of the Nachdiplom lectures formed the basis for these notes. I would also like to thank her for all her help and useful discussions. I would like to thank Michael Struwe, who invited me to give the Nachdiplom course, and Tristan Rivière, Director of the FIM, for inviting me to the Institute (and in particular for inviting the extra guests I suggested while I was there) and I would like to thank Karin Baur for her very kind hospitality in hosting my visit to the ETH and for making my visit so enjoyable. I would like to thank Andrea Waldburger for her help in making the excellent practical arrangements for my visit.

This work was supported by the Engineering and Physical Sciences Research Council [grant number EP/G007497/1], which funds me as a Leadership Fellow at the School of Mathematics at the University of Leeds: I am much indebted to them for the time this has given me to develop my research. I would also like to thank the University of Leeds, and the School of Mathematics, for all their support. I would like to thank Thomas Hintermann and Thomas Kappeler for their help and patience in organising the publication of this book, and the referees for their careful reading of the first version of this manuscript and very useful comments.

This book is dedicated to the memory of Andrei Zelevinsky, who changed a significant part of the mathematical landscape.

Robert Marsh, University of Leeds, August 2013.

Contents

1 Introduction 5

1.1 Motivation for cluster algebras . . . 5
1.2 Some recurrences . . . 7
1.3 Somos recurrences . . . 8
1.4 Why study cluster algebras? . . . 9
1.5 Some notation . . . 9

2 Cluster algebras 10

2.1 Definition of a cluster algebra . . . 10
2.2 Skew-symmetrizable matrices . . . 13
2.3 Quiver notation . . . 13
2.4 Valued quiver notation . . . 16
2.5 Exchange graphs . . . 17

3 Exchange pattern cluster algebras 19

3.1 Exchange patterns . . . 19
3.2 Exchange pattern cluster algebras . . . 22
3.3 Matrices of exponents . . . 23

4 Reflection groups 31

4.1 Definition of a reflection group . . . 31
4.2 Root systems . . . 33
4.3 Simple systems . . . 35
4.4 Coxeter groups . . . 36
4.5 Crystallographic root systems and Cartan matrices . . . 37
4.6 Classification of finite crystallographic reflection groups . . . 39
4.7 Type A_n . . . 39
4.8 Root systems of low rank . . . 41
4.9 Finite Coxeter groups . . . 43
4.10 Reduced expressions . . . 44
4.11 Coxeter elements . . . 47

5 Cluster algebras of finite type 49

5.1 Classification . . . 49
5.2 Folding . . . 53
5.3 Denominators . . . 55
5.4 Root clusters . . . 58
5.5 Admissible sequences of sinks . . . 59
5.6 Number of clusters . . . 62

6 Generalized Associahedra 63
6.1 Fans 63
6.2 The cluster fan 64
6.3 The cluster complex 67
6.4 Normal fans 69
6.5 The generalized associahedron 70
6.6 The generalized associahedron of type A_n 71

7 Periodicity 75
7.1 The Somos-4 recurrence 75
7.2 Period 1 quivers 76
7.3 Periodicity in the coefficient case 78
7.4 Higher period quivers 80
7.5 Categorical periodicity 80

8 Quivers of finite mutation type 84
8.1 Classification 84
8.2 Tagged triangulations 87

9 Grassmannians 90
9.1 Exterior powers 90
9.2 The Grassmannian 91
9.3 The Grassmannian $Gr(2,n)$ 92
9.4 The Grassmannian $Gr(k,n)$ 94
9.5 The Grassmannian $Gr(2,n)$ revisited 98

Bibliography 101

Nomenclature 111

Index 115

1 Introduction

1.1 Motivation for cluster algebras

Cluster algebras are commutative algebras, defined combinatorially by an iterated mutation process. They were introduced by S. Fomin and A. Zelevinsky [67] mainly with the long-term aim of modelling the multiplicative structure of the dual canonical basis associated to the quantized enveloping algebra of a symmetrizable Kac-Moody Lie algebra $\mathfrak{g}$. The fundamental problem they were working on (which is still open), was an explicit description of this basis.

The quantized enveloping algebra $U_q(\mathfrak{g})$ was introduced by V. G. Drinfel'd [44] and M. Jimbo [101], and can be regarded as the q-analogue of the corresponding universal enveloping algebra of $\mathfrak{g}$. G. Lusztig [116, 117, 118] introduced the canonical basis of the quantized enveloping algebra $U_q(\mathfrak{n})$ of the positive part, $\mathfrak{n}$, of $\mathfrak{g}$, in the case where $\mathfrak{g}$ is a symmetric Kac-Moody algebra, and Kashiwara independently introduced the global crystal basis of $U_q(\mathfrak{n})$ in the symmetrizable case. The bases were shown to coincide in the symmetric case in [87], and have beautiful properties, including positivity properties, good representation-theoretic properties and an intriguing combinatorial structure. However, it is very difficult to describe them explicitly. Studying this problem has led to a lot of interesting mathematics.

A complete description of the canonical basis has been given in only a few cases. The bases in types A_1 and A_2 appear already in G. Lusztig's first paper on the canonical basis [116]. Type A_3 is described in [119, 162] and type B_2 in [163]. In type A_4, partial information is known [32, 94, 95, 123].

The canonical basis induces a basis in $\mathbb{C}_q[N]$, the quantum deformation of the algebra $\mathbb{C}[N]$ of regular functions on the pro-unipotent group N associated to $\mathfrak{n}$, which is known as the dual canonical basis, and this basis appears to be more tractable than the canonical basis itself. If N_- denotes the opposite pro-unipotent group to N, set $N(w) = N \cap (w^{-1}N_-w)$ for each element w of the corresponding Weyl group. Then, in [80] it is shown that the coordinate ring $\mathbb{C}[N(w)]$ is a cluster algebra. Furthermore, it is shown that the dual of Lusztig's semicanonical basis (see [121] for the definition) induces bases in each $\mathbb{C}[N(w)]$ containing all cluster monomials: in other words, the cluster structure of $\mathbb{C}[N(w)]$ is relevant to the semicanonical basis, which has a number of key properties in common with the canonical basis. In fact, the semicanonical basis and canonical basis coincide in types A_n for $1 \leq n \leq 4$ [79]. Furthermore, it is shown in [81] that the quantum coordinate ring $\mathbb{C}_q[N(w)]$ is a quantum cluster algebra in the sense of [25] (note that quantum cluster structures have also been studied in a Poisson geometric context in [57, 58, 59]). For another source of information on the dual canonical basis in this context, see [108] and references therein.

Roughly speaking, the idea is to understand the dual canonical basis (of $\mathbb{C}_q[N(w)]$) via its maximal q-commuting subsets. We should start with such a subset, and then try to find more subsets by a process of mutation in which elements are exchanged for new ones. If the process works well, a combinatorial object (such as a quiver, or matrix), can be associated to each q-commuting subset, governing the way in which the mutation process takes place, as well as being mutated itself each time mutation takes place. The set of generators for each q-commuting subset is known as a cluster, and mutation replaces one such generator with a new one. See [81] and references therein for more details concerning this particular quantum cluster algebra structure (see also [22, §6]). There has also been a lot of recent work on bases in cluster algebras per se, e.g. [34, 42, 47, 133, 142].

A second key motivation for the definition of cluster algebras was the notion of total positivity. A real matrix is totally positive if all of its minors are positive, including, for example, all of the entries. Such matrices were studied in a mechanical context in [78]. For an introduction to the relationship between total positivity and cluster algebras, see [61], where the example of G/N, with $G = SL_n(\mathbb{C})$ and N a maximal unipotent subgroup, is discussed in some detail. The coordinate ring $\mathbb{C}[G/N]$ is a cluster algebra. A matrix representing an element in G/N is totally positive if all of its flag minors are positive (i.e., all minors whose column set is an initial set of columns). It is not necessary to check all flag minors are positive in order to check total positivity; instead it is enough to check $(n-1)(n+2)/2$ minors. Categorizing such sets of minors (i.e. total positivity criteria) leads again to the notion of a cluster algebra: such sets behave well under a notion of mutation similar to that described above for the dual canonical basis (and, in fact, this is not a coincidence; for a discussion of the relationship between the canonical basis and total positivity see [120]).

A third key motivation for cluster algebras was the behaviour of certain sequences defined by rational functions which have integer entries. Often such sequences have the property that the nth term is a Laurent polynomial in the initial terms, although a priori it is only a rational function. The integer property then follows, in the case where the initial terms are all set equal to 1. Particularly nice examples of this are the Somos-n sequences (see e.g. [77]) for small n. From this point of view, a cluster algebra should be considered as a kind of recurrence on a regular tree (a tree all of whose vertices have the same valency) in which the formula for the recurrence depends on the vertex of the tree in a well-defined way. The Laurent phenomenon for cluster algebras [67, Thm. 3.1] states that any cluster variable (obtained by arbitrary iterations of the recurrence) is a Laurent polynomial in the initial variables. For more information on the Laurent phenomenon (and, in particular, the Somos sequences from this perspective), see [68]. See also [75] and the recent developments [112, 113].

As well as the fields described above, cluster algebras have been applied to a number of other areas. Cluster categories were introduced in [23] and, independently in type A, in [27], and give a certain kind of categorification of a cluster algebra. A generalization was given in [2]. Cluster categories have had applications to the representation theory of finite dimensional algebras and, in particular, to tilting theory. A more quiver representation-theoretic approach, using the notion of a quiver with potential, is available in [40]. There has been a lot of work in this field, and for more information on this we refer to the surveys [3, 22, 104, 105, 115, 146].

We also refer to two useful computer software packages for cluster algebras, B. Keller's Java applet for quiver mutation [106] and the Sage package on cluster algebras and quivers by G. Musiker and C. Stump [134].

There are a number of relationships between cluster algebras and other areas, and we give references to some of the articles: ad-nilpotent ideals in Borel subalgebras of simple Lie algebras [141], combinatorial geometry (associahedra) [35], discrete integrable systems [75, 103, 136] (and many more), Donaldson-Thomas invariants [107, 109, 110, 135], frieze patterns [5, 7, 8, 14, 15, 19, 88], hyperbolic 3-manifolds [138], BPS quivers and quantum field theories [1], Riemann surfaces [52, 65], Seiberg duality [18, 129, 161] (see also [75, §11]), RNA secondary structure combinatorics [126], shallow water waves and the KP equation [111], Teichmüller theory and Poisson geometry [33, 55, 57, 58, 59, 83, 99], and tropicalization [137]. Space prevents making a list of all references here: there are many more interesting articles on the subject. The list here is meant to give a flavour of the subject, and is not intended to be exhaustive. However, apologies are made for any omissions that should have been included. The *Cluster algebras portal* [60] is a good source of information.

1.2 Some recurrences

Let $k \geq 1$ be an integer, and consider the recurrence defined by $f_1 = x$, $f_2 = y$ and

$$f_{n+1} = \begin{cases} \frac{f_n+1}{f_{n-1}} & \text{if } n \text{ is odd;} \\ \frac{f_n^k+1}{f_{n-1}} & \text{if } n \text{ is even .} \end{cases}$$

If $k = 1$, we obtain the sequence

$$x, y, \frac{y+1}{x}, \frac{x+y+1}{xy}, \frac{x+1}{y}, x, y, \ldots.$$

For $k = 2, 3$, we obtain the sequences:

$$x, y, \frac{y^2+1}{x}, \frac{x+y^2+1}{xy}, \frac{(x+1)^2+y^2}{xy^2}, \frac{x+1}{y}, x, y, \ldots$$

and

$$x, y, \frac{y^3+1}{x}, \frac{x+y^3+1}{xy}, \frac{(x+1)^3+y^3(y^3+2+3x)}{x^2y^3}, \frac{(x+1)^2+y^3}{xy^2},$$
$$\frac{(x+1)^3+y^3}{xy^3}, \frac{x+1}{y}, x, y, \ldots$$

Firstly, by definition, the entries in the sequences are rational functions in x and y, but they turn out to have the stronger property of being Laurent polynomials. Secondly, we obtain sequences of period $5, 6$ and 8, for $k = 1, 2$ and 3 respectively.

Furthermore, the denominators appearing seem to be following a kind of pattern (reminiscent of root systems of rank 2). These properties can be explained by the theory of cluster algebras introduced by S. Fomin and A. Zelevinsky [67]; see Example 2.1.7. These examples turn out to correspond to cluster algebras of finite type: see Chapter 5.

The recurrence with $k = 1$ is known as the *pentagon recurrence*, and can be attributed to R. C. Lyness or N. H. Abel (see [64, §1.1] for more details).

1.3 Somos recurrences

Choose an integer $r \geq 2$, and consider the recurrence:

$$x_n x_{n+r} = \begin{cases} x_{n+1}x_{n+r-1} + x_{n+2}x_{n+r-2} + \dots x_{n+\frac{r}{2}}x_{n-\frac{r}{2}} & \text{if } r \text{ is even;} \\ x_{n+1}x_{n+r-1} + x_{n+2}x_{n+r-2} + \dots x_{n+\frac{r-1}{2}}x_{n+\frac{r+1}{2}} & \text{if } r \text{ is odd,} \end{cases}$$

with initial conditions $x_1 = \dots = x_r = 1$. This is known as the *Somos-r recurrence*. For example, if $r = 4$, the recurrence is defined by

$$x_n x_{n+3} = x_{n+1}x_{n+3} + x_{n+2}^2.$$

The first terms of the sequence are:

$$1, 1, 1, 1, 2, 3, 7, 23, 59, 314, 1529, 8209, \dots.$$

It is immediate from the definition that the Somos-r sequence consists of rational numbers. If $r = 4, 5, 6$ or 7 then it is known that the entries are all integers. For $r = 4$ or 5, this was proved independently by a number of authors in 1990, including J. L. Malouf [122, §1], Enrico Bombieri, and Dean Hickerson, according to J. Propp [145]. See also [77]. Propp also mentions that the cases $r = 6$ and $r = 7$ were proved by Dean Hickerson and Ben Lotto respectively in 1990.

We can treat the first r terms of the Somos-r sequence as indeterminates. Then, it was shown by S. Fomin and A. Zelevinsky [68], for $r = 4, 5, 6, 7$, that the nth term of the Somos-r sequence can be written as a Laurent polynomial in the first r terms. The proof used the theory of cluster algebras. It follows from this that the entries in the sequence, with initial conditions as above, are integers, giving an independent proof of this integrality.

According to entry A030127 in [140], the Somos-r sequences for for $r \geq 8$ are not integral, e.g. $x_{17} \notin \mathbb{Z}$ for Somos-8. This entry is the list of first terms in Somos-r sequences which are non-integral for $r \geq 8$. There are also interesting connections with elliptic curves: see [26, 50].

1.4 Why study cluster algebras?

Here we list some reasons for studying cluster algebras.

- An interest in the representation theory of quivers and finite dimensional algebras: cluster algebras give new families of algebras and new ways to compare categories (cluster-tilting theory).
- An interest in root systems and Weyl groups. For example, cluster algebras of finite type correspond to crystallographic root systems, and have a Dynkin classification.
- An interest in discrete integrable systems. In a certain sense, cluster algebras can be considered as such systems and there are interesting integrals on related recurrences.
- An interest in combinatorics. There is some beautiful combinatorics associated to cluster algebras, e.g. via root systems or representation theory, including generalized associahedra. The type A associahedron arises from the associativity rule, and has vertices corresponding to different ways of bracketing a sum or product.
- An interest in Riemann surfaces and Teichmüller theory. Many cluster algebras have a combinatorial description in terms of the geometry of surfaces, giving new perspectives on the representation theory mentioned above.

1.5 Some notation

Throughout the book, we will use the following notation, for real numbers x and integers a, b with $a \leq b$.

$$\begin{aligned}
[x]_+ &= \max(x, 0); \\
[x]_- &= \min(0, x); \\
\operatorname{sgn}(x) &= \begin{cases} 1 & \text{if } x > 0; \\ 0 & \text{if } x = 0; \\ -1 & \text{if } x < 0; \end{cases} \\
[a, b] &= \{a, a+1, \ldots, b\}.
\end{aligned}$$

2 Cluster algebras

In this chapter we give the definition and first properties of cluster algebras, including exchange graphs. We mainly follow the definitions in [67], noting also later development in [17, 70]. We will focus in these notes on cluster algebras of geometric type, although more general versions have been defined.

2.1 Definition of a cluster algebra

Fix $n, m \in \mathbb{N}$ satisfying $n \leq m$. We'll usually denote the set $\{1, 2, \ldots, n\}$ by I. Let $\mathbb{F} = \mathbb{Q}(u_1, \ldots, u_m)$ be the field of rational functions in the indeterminates $u_1, \ldots, u_m$ with coefficients in $\mathbb{Q}$. Thus the elements of $\mathbb{F}$ are of the form $\frac{f(u_1,\ldots,u_m)}{g(u_1,\ldots,u_m)}$ where f, g are polynomials with coefficients in $\mathbb{Z}$. A cluster algebra (of geometric type) is a subring of $\mathbb{F}$ determined by combinatorial data. The data determine a subset of $\mathbb{F}$ which generates the cluster algebra.

Definition 2.1.1. [67, 70] A *seed* is a pair $(\widetilde{\mathbf{x}}, \widetilde{B})$ where:

(a) $\widetilde{\mathbf{x}} = \{x_1, \ldots, x_m\}$ is a free generating set of $\mathbb{F}$ over $\mathbb{Q}$, i.e. $\mathbb{F}$ is generated as a field over $\mathbb{Q}$ by $\widetilde{\mathbf{x}}$, and the x_i for $1 \leq i \leq m$ are algebraically independent.

(b) $\widetilde{B} = (b_{ij})_{1 \leq i \leq m, 1 \leq j \leq n}$ is an $m \times n$ integer matrix where the first n rows and the columns correspond to the elements of $\mathbf{x} = \{x_1, \ldots, x_n\}$ and the last $m - n$ rows correspond to the elements of $\mathbf{c} = \{x_{n+1}, \ldots, x_m\}$.

(c) The *principal part* $B = (b_{ij})_{1 \leq i,j \leq n}$ of $\widetilde{B}$ is sign-skew-symmetric, i.e. for all $1 \leq i, j \leq n$, either $b_{ij} = b_{ji} = 0$, or b_{ij} and b_{ji} are non-zero and of opposite sign. In particular, b_{ii} must be zero for all i.

The set $\mathbf{x}$ is known as the *cluster* of the seed, with its elements known as *cluster variables*. The elements of the set $\mathbf{c}$ are known as *coefficients*, *stable variables* or *frozen variables*, and the union $\widetilde{\mathbf{x}} = \mathbf{x} \cup \mathbf{c}$ is known as the *extended cluster* of the seed. The matrix $\widetilde{B}$ is known as an *exchange matrix*. If $m = 2n$ and the lower part $(b_{ij})_{n+1 \leq i \leq 2n, 1 \leq j \leq n}$ of the matrix $\widetilde{B}$ is the identity matrix we say that we are in the case of *principal coefficients* (see [72, Defn. 3.1]).

Definition 2.1.2. [67, §7] (see also [17, §1]) Two seeds $(\widetilde{\mathbf{x}}, \widetilde{B})$ and $(\widetilde{\mathbf{y}}, \widetilde{C})$ are said to be *equivalent* if there is a permutation π of $[1, m]$ such that

(a) $\pi(I) = I$;

(b) $\pi(i) = i$ for all $i \in \{n + 1, \ldots, m\}$;

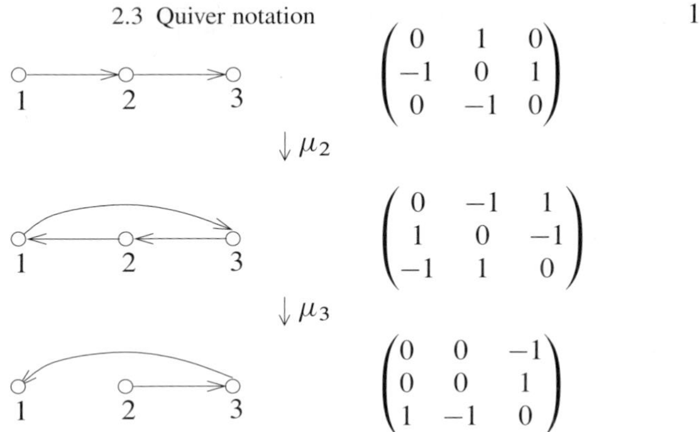

Figure 2.1. Example of quiver and matrix mutation.

We mutate it at vertex 1. Step 1 in the mutation rule yields:

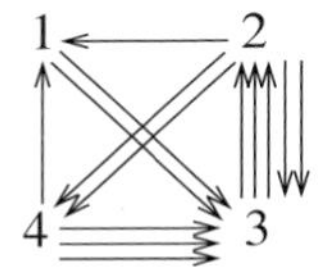

Step 2 yields:

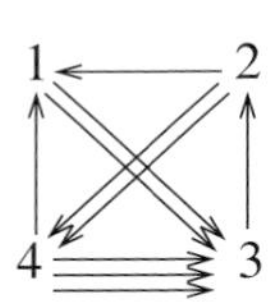

Step 3 yields the final mutated quiver, $\mu_1 S_4$:

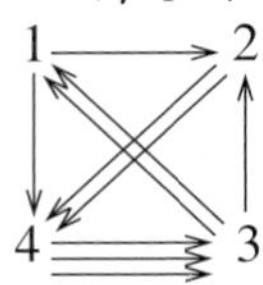

Notice that $\mu_1 S_4$ is the original quiver, S_4, rotated through an angle of $\frac{\pi}{2}$ anticlockwise (with a different numbering) — we shall return to this later.

A very useful device for mutating quivers is B. Keller's Java applet [106].

Note that the exchange relation (equation (2.1)) also looks very nice in the quiver notation. Fixing a vertex k, we have:

$$x_k x_k' = \prod_{i \to k} x_i + \prod_{k \to i} x_i$$

where the products are taken over $1 \leq i \leq m$, again with multiplicities taken into account. So, one can define a cluster algebra using a *quiver seed* $(\widetilde{x}, \widetilde{Q})$ (where $\widetilde{Q} = Q(\widetilde{B})$).

2.4 Valued quiver notation

A *valued graph* G [43] on vertices $1, \dots, n$ is a graph with no loops and at most one arrow between any pair of vertices, together with integers v_{ij} for all i, j for which there is an edge between i and j, such that there are positive integers $d_1, \dots, d_n$ such that $d_i v_{ij} = d_j v_{ji}$ for all i, j, i.e. the matrix (v_{ij}) (with zero entries where there is no edge) is symmetrizable. If $v_{ij} = v_{ji} = 1$ for all i, j, then Q is regarded as a graph in the usual sense. We write v_{ij} on the edge between i and j near i and v_{ji} near j. For edges with $v_{ij} = v_{ji} = 1$, we omit the labels.

An orientation of a valued graph is just an orientation of its underlying graph, and is known as a *valued quiver*. Given a skew-symmetrizable matrix (b_{ij}) with $d_i b_{ij} = -d_j b_{ji}$ for all i, j we can define a valued quiver on vertices $1, \dots, n$ by the following. If $b_{ij} \neq 0$ (and therefore $b_{ji} \neq 0$), set

$$v_{ij} = |b_{ij}|, \qquad v_{ji} = |b_{ji}|$$

and put an arrow in the quiver from i to j if $b_{ij} > 0$ or from j to i if $b_{ij} < 0$. It is clear that this gives a bijection between skew-symmetrizable matrices and valued quivers. Mutation of skew-symmetrizable matrices (see Definition 2.1.3) induces a mutation of valued quivers. Thus we can also work with *valued quiver seeds* $(\mathbf{x}, Q)$, where Q is a valued quiver, in the skew-symmetrizable case.

Remark 2.4.1. There are slightly different ways of thinking about valued quivers. For example, in [103, §9.1], a valued quiver is defined as a quiver Q with no loops and at most one arrow between any pair of vertices (as above), together with a valuation map $v : Q_1 \to \mathbb{N}^2_{>0}$ taking each arrow α to a pair $(v(\alpha)_1, v(\alpha)_2)$ of positive integers, and a map $d : Q_0 \to \mathbb{N}_{>0}$ such that $v(\alpha)_1 d(i) = v(\alpha)_2 d_j$ whenever $\alpha : i \to j$ is an arrow in Q. If Q is a valued quiver in the sense defined above, then setting $v(\alpha) = (v_{ij}, v_{ji})$ whenever $\alpha : i \to j$ is an arrow in Q_1 makes Q into a valued quiver as considered in [103, §9.1].

As an example, we consider Figure 2.2, where the mutation of a valued quiver (at vertex 3) is shown. Note that the valued arrow α is reversed, to give a new arrow α^* in the mutated quiver. In the setting of [103, §9.1] we have $v(\alpha) = (1, 2)$ while $v(\alpha^*) = (2, 1)$.

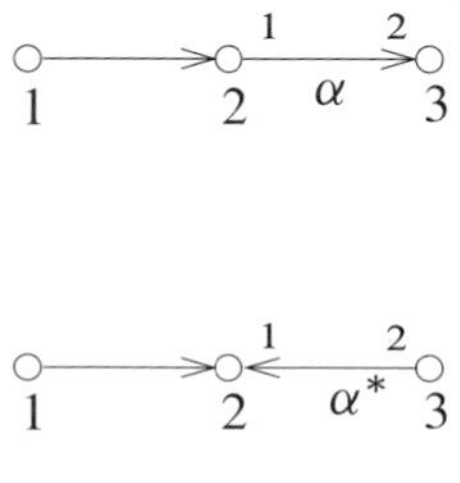

$$\underset{1}{\circ} \longrightarrow \underset{2}{\circ} \xrightarrow[\alpha]{1 \qquad 2} \underset{3}{\circ} \qquad \begin{pmatrix} 0 & 1 & 0 \\ -1 & 0 & 1 \\ 0 & -2 & 0 \end{pmatrix}$$

$$\downarrow \mu_3$$

$$\underset{1}{\circ} \longrightarrow \underset{2}{\circ} \xleftarrow[\alpha^*]{1 \qquad 2} \underset{3}{\circ} \qquad \begin{pmatrix} 0 & 1 & 0 \\ -1 & 0 & -1 \\ 0 & 2 & 0 \end{pmatrix}$$

Figure 2.2. Mutation of a valued quiver of type B_3.

If the cluster algebra has coefficients, these can be encoded by adding arrows to the quiver between frozen vertices i and vertices $j \in I$. If $b_{ij} > 0$, put an arrow from i to j with a valuation $v_{ij} = |b_{ij}|$ near i only. If $b_{ij} < 0$, put an arrow from j to i with a valuation $v_{ij} = |b_{ij}|$ near i only.

2.5 Exchange graphs

Definition 2.5.1. [67, §7] The *exchange graph* of a cluster algebra is defined as follows:

$$\text{Vertices} \longleftrightarrow \text{seeds}$$
$$\text{Edges} \longleftrightarrow \text{mutations.}$$

Thus the exchange graph of a cluster algebra of rank n is n-regular, i.e. each vertex has valency n.

Example 2.5.2. The exchange graph of $\mathcal{A}(0)$ has two vertices, corresponding to the two seeds $((x_1), (0))$ and $(2/x_1, (0))$, with a single edge between them.

Let $n = 2$ and $m = 4$, $\widetilde{B} = \begin{pmatrix} 0 & 1 \\ -1 & 0 \\ 1 & 0 \\ 0 & 1 \end{pmatrix}$ and $\widetilde{\mathbf{x}} = \{x_1, x_2, x_3, x_4\}$. Then the exchange graph is as shown in Figure 2.3. Note that for the mutation corresponding to the top left edge of the graph (marked with an asterisk), the equivalence rule of Definition 2.1.2 must be applied to see that the mutated seed is as claimed.

Note that if x_3 and x_4 are deleted, and the bottom two rows of $\widetilde{B}$ are deleted (i.e. we only take the principal part of $\widetilde{B}$), then we obtain an isomorphic graph (but with different labels). The cluster variables in this case can be obtained from those in the figure by setting x_3 and x_4 to 1. This cluster algebra corresponds to the matrix $B = \begin{pmatrix} 0 & 1 \\ -1 & 0 \end{pmatrix}$, considered in Example 2.1.7 (for $k = 1$).

Definition 2.5.3. [70, §1] Let $\mathcal{A}(\widetilde{\mathbf{x}}, \widetilde{B}) \subseteq \mathbb{F}$ and $\mathcal{A}(\widetilde{\mathbf{y}}, \widetilde{C}) \subseteq \mathbb{F}'$ be cluster algebras. We say that $\mathcal{A}(\widetilde{\mathbf{x}}, \widetilde{B})$ and $\mathcal{A}(\widetilde{\mathbf{y}}, \widetilde{C})$ are *strongly isomorphic* (or isomorphic as cluster algebras) if there is a field isomorphism $\sigma : \mathbb{F} \to \mathbb{F}'$ such that $(\sigma(\widetilde{\mathbf{x}}), \widetilde{B})$ is a seed in $\mathcal{A}(\widetilde{\mathbf{y}}, \widetilde{C})$; i.e. if and only if there is a seed in $\mathcal{A}(\widetilde{\mathbf{y}}, \widetilde{C})$ whose exchange matrix is $\widetilde{B}$. Note that the exchange graphs of strongly isomorphic cluster algebras are necessarily isomorphic.

In Example 2.5.2, the number of seeds turned out to be finite — this does not always happen; e.g.

$$\begin{pmatrix} 0 & 2 \\ -2 & 0 \end{pmatrix}$$

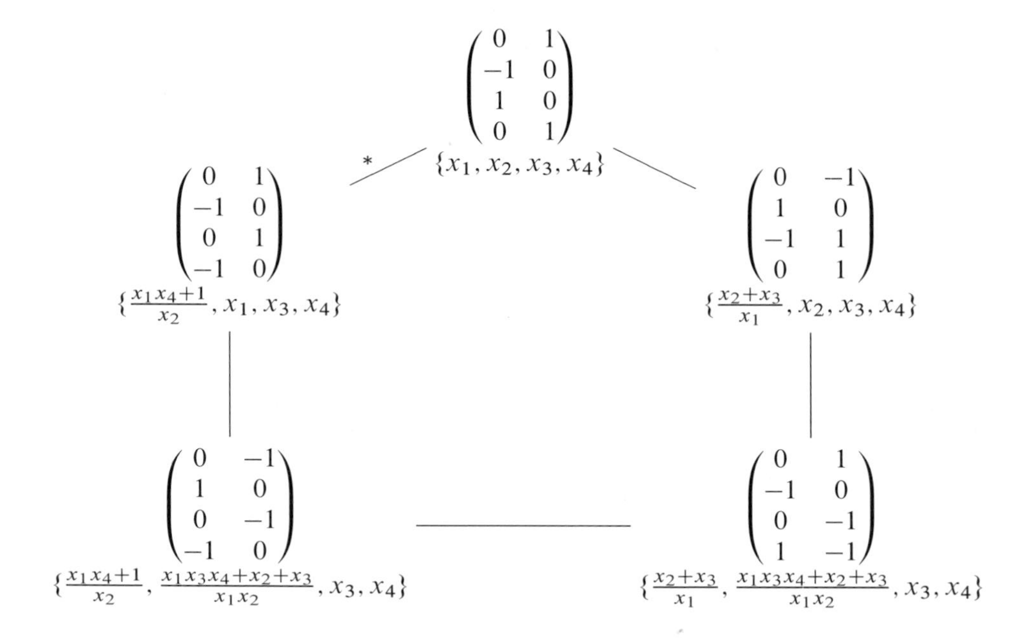

Figure 2.3. An example of an exchange graph. The asterisk denotes a mutation where a non trivial seed equivalence arises.

gives infinitely many seeds. A cluster algebra is said to be of *finite type* if it has finitely many seeds. In order to study the classification we first need to study reflection groups and (crystallographic) root systems. Before that, we consider an alternative definition of a cluster algebra.

3 Exchange pattern cluster algebras

The first definition of a cluster algebra (also from [67]) involved the polynomials more directly, with mutation being defined by substitutions rather than via sign-skew-symmetric matrices. In this chapter we describe this approach in detail and explain how it corresponds to the version with matrices given in the previous chapter, following [67]. This version was used in the proof of the Laurent phenomenon (which will be discussed later; see Theorem 7.1.1), and was also used in recent work of T. Lam and P. Pylyavskyy [112, 113] generalizing cluster algebras.

3.1 Exchange patterns

Recall that $I = \{1, 2, \dots, n\}$. Let $\mathbb{T}_n$ be an n-regular tree, i.e. an acyclic graph in which each vertex has valency n. We label the edges in such a way that the edges incident with any given vertex are labelled $1, 2, \dots, n$. Examples for the cases $n = 1, 2, 3$ are given in Figure 3.1. We also use $\mathbb{T}_n$ to denote the vertices of $\mathbb{T}_n$.

Let $\mathbb{P}$ denote a torsion-free abelian group, written multiplicatively. This is known as the *coefficient group*. For example, $\mathbb{P}$ could be a free abelian group of finite rank.

To each vertex $t \in \mathbb{T}_n$ we associate a *cluster* $\mathbf{x}(t) = \{x_i(t)\}_{i \in I}$ of n commuting *cluster variables*. To each $t \in \mathbb{T}_n$ and $j \in I$, we associate a monomial

$$M_j(t) = M_j(\mathbf{x}(t)) = p_j(t) \prod_{i \in I} x_i(t)^{a_i}$$

where the $p_j(t)$ lie in $\mathbb{P}$, and the a_i are nonnegative integers. We regard $M_j(t)$ as labelling the end of the edge incident with $t \in \mathbb{T}_n$ labelled with $j \in I$. The cluster variables are specified to satisfy the *exchange relations*, i.e. for all edges in $\mathbb{T}_n$ as follows:

$$t \underset{M_j(t) \qquad M_j(t')}{\overset{j}{\text{———————}}} t'$$

we have:

$$x_i(t) = x_i(t') \text{ for all } i \neq j \tag{3.1}$$

$$x_j(t)x_j(t') = M_j(x(t)) + M_j(x(t')) \tag{3.2}$$

We often refer to (3.2) as the exchange relation, and the right hand side of this equation as the *exchange polynomial*.

Definition 3.1.1. [67, §2] We say that the monomials $\mathcal{M} = (M_j(t))$ form an *exchange pattern* (on $\mathbb{T}_n$) if the following hold:

$n = 1$ 1

$n = 2$ 2 1 2 1

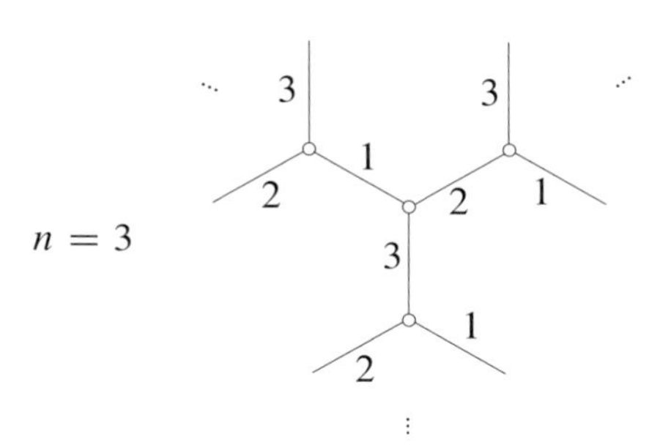

Figure 3.1. Regular trees of rank 1, 2 and 3.

(E1) For all $t \in \mathbb{T}_n$, $x_j(t) \nmid M_j(t)$.

(E2) If $t_1 \overset{}{\underset{j}{\text{———}}} t_2$ and $x_i(t_1) | M_j(t_1)$ then $x_i(t_2) \nmid M_j(t_2)$.

(E3) If $t_1 \overset{}{\underset{j}{\text{———}}} t_2$, and $i \neq j$, then

$$x_i(t_1) | M_j(t_1) \text{ if and only if } x_j(t_2) | M_i(t_2).$$

(E4) If $t_1 \overset{}{\underset{j}{\text{———}}} t_2 \overset{}{\underset{k}{\text{———}}} t_3 \overset{}{\underset{j}{\text{———}}} t_4$, with $j \neq k$, then:

$$\frac{M_j(t_3)}{M_j(t_4)} = \left. \frac{M_j(t_2)}{M_j(t_1)} \right|_{x_k \leftarrow M_0 / x_k}.$$

Here, x_k represents $x_k(t_1) = x_k(t_2)$, and

$$M_0 = (M_k(t_2) + M_k(t_3))|_{x_j = 0}$$

(with $x_j = x_j(t_2) = x_j(t_3)$).

Remark 3.1.2. (a) By (E1), $x_j(t) \nmid M_j(t)$, so the exchange polynomial

$$P = P(\mathbf{x}(t)) = M_j(\mathbf{x}(t)) + M_j(\mathbf{x}(t'))$$

is actually a function of $\mathbf{x}(t)$ (or of $\mathbf{x}(t')$) by (3.1).

(b) In (E4), M_0 is monomial whenever it appears in the substitution. For, suppose that $x_j(t_2) \nmid M_k(t_2)$ and $x_j(t_3) \nmid M_k(t_3)$, i.e. M_0 is not monomial. Applying (E3) to $t_1 \overset{}{\underset{j}{\text{———}}} t_2$, we see that, since $x_j(t_2) \nmid M_k(t_2)$, we have $x_k(t_1) \nmid M_j(t_1)$. Applying (E3) to $t_3 \underset{k}{\text{———}} t_2$, we see that, since $x_j(t_3) \nmid M_k(t_3)$, we have $x_k(t_2) \nmid M_j(t_2)$. Since x_k does not appear in $M_j(t_1)$ or $M_j(t_2)$, we see that M_0 is not substituted in (E4).

(c) Suppose we apply (E4) to the reverse sequence of edges, i.e. to

$$t_4 \underset{j}{\text{———}} t_3 \underset{k}{\text{———}} t_2 \underset{j}{\text{———}} t_1 .$$

We obtain

$$\frac{M_j(t_2)}{M_j(t_1)} = \frac{M_j(t_3)}{M_j(t_4)}|_{x_k \leftarrow M_0/x_k},$$

where

$$M_0 = M_k(t_2) + M_k(t_3)|_{x_j=0},$$

as before. Making the substitution $x_k \leftarrow M_0/x_k$ on both sides, we obtain

$$\frac{M_j(t_2)}{M_j(t_1)}|_{x_k \leftarrow M_0/x_k} = \frac{M_j(t_3)}{M_j(t_4)}|_{x_k \leftarrow M_0/(M_0/x_k)} = \frac{M_j(t_3)}{M_j(t_4)}.$$

Thus, we only obtain the same conclusion obtained by applying (E4) to the original sequence of edges.

(d) There is a natural involution on exchange patterns. Suppose that $(M_j(t))$ is an exchange pattern. Define $M'_j(t) = M_j(t')$ for every edge $t \underset{j}{\text{———}} t'$. Then it is easy to check that $\mathcal{M}' = (M'_j(t))$ is also an exchange pattern.

(e) [67, Rk. 2.2] Axiom (E4) can be considered as describing the 'propagation' of an exchange pattern along the edges of $\mathbb{T}_n$. Fix a vertex t of $\mathbb{T}_n$ and let $t \underset{j}{\text{———}} t(j)$, $j \in I$, be the edges incident with t. Suppose that $M_j(t)$ and $M_j(t(j))$ are known for all j. Let $j, k \in I$ with $j \neq k$ and consider the edges:

$$t(j) \underset{j}{\text{———}} t \underset{k}{\text{———}} t(k) \underset{j}{\text{———}} t' .$$

Applying (E4) with $t_1 = t(j)$, $t_2 = t$, $t_3 = t(k)$ and $t_4 = t'$, we have

$$\frac{M_j(t(k))}{M_j(t')} = \left.\frac{M_j(t)}{M_j(t(j))}\right|_{x_k \leftarrow M_0/x_k},$$

where

$$M_0 = (M_k(t) + M_k(t(k)))|_{x_j=0}.$$

Hence, $M_j(t(k))/M_j(t')$ is determined. By (E2), this uniquely determines the exponents in $M_j(t(k))$ and $M_j(t')$. It also determines the ratio $p_j(t')/p_j(t'')$

but not the individual terms in this fraction — there is still some degree of freedom. Allowing j to vary, we are able to compute the same information for $t(k)$ that we had for t (noting also that $M_k(t(k))$ and $M_k(t)$ are already given). The freedom in the coefficients disappears for so-called normalized exchange patterns; see [67, §5].

3.2 Exchange pattern cluster algebras

Let $\mathbb{P}$ be a torsion-free abelian group and $\mathcal{M} = (M_j(t))_{t\in\mathbb{T}_n}$ be an exchange pattern, with $M_j(t) \in \mathbb{ZP}[x_j(t) : j \in I]$ for $t \in \mathbb{T}_n$ and $j \in I$. Since $\mathbb{P}$ is torsion-free, $\mathbb{ZP}$ has no zero-divisors. For $t \in \mathbb{T}_n$, let $\mathbb{F}(t)$ be the field of rational functions in the $x_j(t)$, with coefficients in $\mathbb{ZP}$. This is the field of fractions of the integral domain $\mathbb{ZP}[x_j(t) : j \in I]$.

For an edge $t \overset{}{\underset{k}{\text{———}}} t'$ in $\mathbb{T}_n$, let $P_{t,t'} : \mathbb{F}(t') \to \mathbb{F}(t)$ be the map taking $x_i(t')$ to $x_i(t)$ for $i \neq k$ and $x_k(t')$ to $P(\mathbf{x}(t))/x_k(t)$. This is a $\mathbb{ZP}$-linear field isomorphism with inverse $P_{t',t}$.

In this way we can identify all the fields $\mathbb{F}(t)$, for $t \in \mathbb{T}_n$, with each other, as a field $\mathbb{F}$ containing all $x_i(t)$, $t \in \mathbb{T}_n$, $i \in I$, elements satisfying the exchange relations (3.1) and (3.2) above.

Definition 3.2.1. (*Exchange pattern cluster algebra*) [67, §2] Let $\mathbb{A}$ be a subring of $\mathbb{ZP}$ with 1 containing all of the elements $p_i(t)$, for $t \in \mathbb{T}_n$ and $i \in I$. Thus, for example, $\mathbb{A}$ could be the smallest such subring, or $\mathbb{ZP}$ itself. The exchange pattern cluster algebra $\mathcal{A} = \mathcal{A}_{\mathbb{A}}(\mathcal{M})$ is the $\mathbb{A}$-subalgebra with 1 of $\mathbb{F}$ generated by

$$\bigcup_{t\in\mathbb{T}_n} \mathbf{x}(t).$$

It is said to be the exchange pattern cluster algebra of rank n over $\mathbb{A}$ associated to the exchange pattern $\mathcal{M}$.

Since $\mathbb{F}$ is a field, $\mathcal{A}$ is an integral domain; it has no zero-divisors.

Remark 3.2.2. It is clear that if $\mathcal{M}'$ is defined as in Remark 3.1.2(d) then

$$\mathcal{A}_{\mathbb{A}}(\mathcal{M}) = \mathcal{A}_{\mathbb{A}}(\mathcal{M}').$$

3.3 Matrices of exponents

We first of all assume that $p_j(t) = 1$ for all $j \in I$ and $t \in \mathbb{T}_n$.

Lemma 3.3.1. *[67, §4] A collection of monomials $M_j(t) = \prod_{i \in I} x_i(t)^{a_i}$ satisfies (E1) and (E2) if and only if there are $n \times n$ integer matrices $B(t) = (b_{ij}(t))$, $t \in \mathbb{T}_n$, satisfying*

$$b_{jj}(t) = 0 \text{ for all } j \in I,\ t \in \mathbb{T}_n \tag{3.3}$$

and

$$b_{ij}(t') = -b_{ij}(t) \tag{3.4}$$

whenever $t \overset{}{\underset{j}{\text{———}}} t'$ is an edge in $\mathbb{T}_n$ and $i \in I$, $i \neq j$, and

$$M_j(t) = \prod_{i \in I} x_i(t)^{[b_{ij}(t)]_+} \text{ for all } j \in I,\ t \in \mathbb{T}_n. \tag{3.5}$$

Proof. Given a set of monomials $M_j(t)$ satisfying (E1) and (E2), note that by (E1), $x_j(t) \nmid M_j(t)$ and $x_j(t') \nmid M_j(t')$. It follows that we can write:

$$\frac{M_j(t)}{M_j(t')} = \prod_{i \in I, i \neq j} x_i^{b_{ij}(t)}$$

whenever $t \overset{}{\underset{j}{\text{———}}} t'$ is an edge in $\mathbb{T}_n$, where the $b_{ij}(t)$ are integers. Here $x_i = x_i(t) = x_i(t')$ for all $i \in I, i \neq j$. We set $b_{jj}(t) = 0$ for all $j \in I$. Then $B(t) = (b_{ij}(t))_{i,j \in I}$ is an $n \times n$ integer matrix satisfying (3.3). By (E2), no x_i divides both $M_j(t)$ and $M_j(t')$, and (3.5) follows.

Conversely, given a collection $B(t) = (b_{ij}(t))_{i,j=1}^n$, $t \in \mathbb{T}_n$, of integer matrices satisfying (3.3) and (3.4), we may define monomials $M_j(t)$ using (3.5). It follows from this definition that (E1) holds. To see (E2), suppose that $t \overset{}{\underset{j}{\text{———}}} t'$ is an edge in $\mathbb{T}_n$, that $i \neq j$ and that $x_i | M_j(t_1)$ (where $x_i = x_i(t_1) = x_i(t_2)$). Then $b_{ij}(t_1) > 0$, so $b_{ij}(t_2) < 0$ by (3.4). Then

$$M_j(t_2) = \prod_{i \in I} x_i^{[b_{ij}(t_2)]_+},$$

so $x_i \nmid M_j(t_2)$ as required. □

We remark that if the $M_j(t)$ satisfy the conditions in Lemma 3.3.1, then, whenever $t \overset{}{\underset{j}{\text{———}}} t'$ is an edge in $\mathbb{T}_n$, we have:

$$M_j(t') = \prod_{i \in I} x_i(t)^{-[b_{ij}(t)]_-}.$$

Lemma 3.3.2. *[67, Proof of Prop. 4.3] Suppose that the monomials $M_j(t)$ satisfy the conditions in Lemma 3.3.1. Then the $M_j(t)$ satisfy (E3) if and only if the following holds:*

$$b_{ij}(t) > 0 \text{ if and only if } b_{ji}(t) < 0, \text{ for all } i \neq j \in I \text{ and } t \in \mathbb{T}_n.$$

Proof. Recall that (E3) states that for any edge $t_1 \overset{}{\underset{j}{\text{———}}} t_2$ in $\mathbb{T}_n$, we have (for $i \neq j$):

$$x_i(t_1)|M_j(t_1) \text{ if and only if } x_j(t_2)|M_i(t_2),$$

i.e.

$$b_{ij}(t_1) > 0 \text{ if and only if } b_{ji}(t_2) > 0,$$

i.e.

$$-b_{ij}(t_2) > 0 \text{ if and only if } b_{ji}(t_2) > 0$$

(using (3.4)), i.e.

$$b_{ij}(t_2) < 0 \text{ if and only if } b_{ji}(t_2) > 0.$$

The result follows. □

Lemma 3.3.3. *[67, Proof of Prop. 4.3] Suppose that the monomials $M_j(t)$ satisfy the conditions in Lemma 3.3.1 and also satisfy (E3). Then the $M_j(t)$ satisfy (E4) if and only if, for any edge $t \underset{k}{\text{———}} t'$ in $\mathbb{T}_n$, we have:*

$$b_{ij}(t') = \begin{cases} -b_{kj}(t) & \text{if } i = k; \\ b_{ij}(t) + sgn(b_{ik})[b_{ik}b_{kj}]_+ & \text{if } i \neq k \text{ and } j \neq k. \end{cases} \tag{3.6}$$

Proof. We first remark that the case $j = k$ is dealt with by (3.4) (i.e. $b_{ik}(t') = -b_{ik}(t)$). Note that (E4) states that, for all subgraphs

$$t_1 \underset{j}{\text{———}} t_2 \underset{k}{\text{———}} t_3 \underset{j}{\text{———}} t_4$$

in $\mathbb{T}_n$, we have (using E1):

$$\prod_{i \in I} x_i^{b'_{ij}} = \prod_{i \in I} x_i^{b_{ij}} \Bigg|_{x_k \leftarrow M_0/x_k}, \tag{3.7}$$

where

$$M_0 = \left(\prod_{i \in I} x_i^{[b_{ik}]_+} + \prod_{i \in I} x_i^{-[b_{ik}]_-} \right) \Bigg|_{x_j = 0},$$

where, for $i \neq k$, we write $x_i = x_i(t_2) = x_i(t_3)$, $b_{ij} = b_{ij}(t_2)$ and $b'_{ij} = b_{ij}(t_3)$. We compute the exponent of each x_i on the right hand side of (3.7). Firstly, we consider the case $i = k$. If $b_{kj} \neq 0$, the exponent of x_k is $-b_{kj}$, since x_k does not appear in

since $c'_{ik} = -c_{ik}$. Hence, the $p_j(t)$ form an exchange pattern if and only if, for all $i \in I'$ and $j \in I$, $j \neq k$,

$$c'_{ij} = c_{ij} + \operatorname{sgn}(c_{ik})[c_{ik}b_{kj}]_+$$

and we are done. □

We can now put all these results together to obtain:

Theorem 3.3.8. *[67, §5] Suppose that $\mathbb{P}$ is a free abelian group on generators x_i, $i \in I'$, a finite set. For each $t \in \mathbb{T}_n$ and $j \in I$, suppose we have a monomial*

$$M_j(t) = p_j(t) \prod_{i \in I} x_i(t)^{a_i},$$

where the $p_j(t)$ lie in $\mathbb{P}$ and the a_i are nonnegative integers. Then the monomials $M_j(t)$, for $t \in T$, $j \in I$, form an exchange pattern of geometric type if and only if there is a collection of matrices $\widetilde{B}(t)$, for $t \in \mathbb{T}_n$, with rows indexed by $I \cup I'$ and columns indexed by I, satisfying:

(a) The matrix $B(t)$ (obtained by taking the rows of $\widetilde{B}(t)$ indexed by I only) is sign-skew-symmetric for all $t \in \mathbb{T}_n$.

(b) Whenever $t \overset{}{\underset{k}{\text{———}}} t'$ is an edge in $\mathbb{T}_n$, we have $\widetilde{B}(t') = \mu_k(\widetilde{B}(t))$.

and, for all $j \in I$ and $t \in \mathbb{T}_n$,

$$M_j(t) = \prod_{i \in I \cup I'} x_i(t)^{[b_{ij}(t)]_+}.$$

Proof. Suppose first that the monomials $M_j(t)$ form an exchange pattern of geometric type. Let $\overline{M}_j(t)$ denote the same monomials with the coefficients specialised to 1. Then the $\overline{M}_j(t)$ also satisfy (E1)-(E4) in Definition 3.1.1. By Proposition 3.3.4, there is a family $B(t)$, $t \in \mathbb{T}_n$ of sign-skew-symmetric $n \times n$ integer matrices such that whenever $t \underset{k}{\text{———}} t'$ is an edge in $\mathbb{T}_n$, we have $B(t') = \mu_k(B(t))$ and, for all $j \in I$ and $t \in \mathbb{T}_n$,

$$\overline{M}_j(t) = \prod_{i \in I} x_i(t)^{[b_{ij}(t)]_+}.$$

By Proposition 3.3.7, the $p_j(t)$ have the form

$$p_j(t) = \prod_{i \in I'} x_i^{[c_{ij}(t)]_+},$$

where $C(t) = (c_{ij}(t))_{i \in I', j \in I}$ are matrices satisfying:
For any edge $t \underset{k}{\text{———}} t'$ in $\mathbb{T}_n$, $C(t) = (c_{ij}(t))$ and $C(t') = (c_{ij}(t'))$ are related by the equation:

$$c'_{ij}(t) = \begin{cases} -c_{ij}(t) & \text{if } j = k; \\ c_{ij}(t) + \operatorname{sgn}(c_{ik})[c_{ik}b_{kj}]_+ & \text{otherwise.} \end{cases}$$

Defining $\widetilde{B}(t)$ to be the matrix with rows indexed by $I \cup I'$ and columns indexed by I, with the entries in the rows indexed by I coinciding with the entries of B and the entries in the rows indexed by I' coinciding with the entries of C, we see that the $M_j(t)$ satisfy the conditions in the theorem.

Conversely, suppose the $M_j(t)$ are as described in the theorem. Then the $\overline{M}_j(t)$ are as described in Proposition 3.3.4, and thus form an exchange pattern by that result, and it follows from Proposition 3.3.7 that the $M_j(t)$ form an exchange pattern of geometric type, as required. □

An exchange pattern cluster algebra is said to be of *geometric type* if it is defined by an exchange pattern of geometric type as in Theorem 3.3.8.

Theorem 3.3.9. *[67, §5] A commutative ring is a cluster algebra of geometric type if and only if it is an exchange pattern cluster algebra of geometric type (as in Theorem 3.3.8) such that the ring $\mathbb{A}$ in the definition (Definition 3.2.1) is the subring of $\mathbb{ZP}$ generated by the elements x_i, $i \in I'$.*

Proof. Let $\mathcal{A}(\widetilde{\mathbf{x}}, \widetilde{B}) \subseteq \mathbb{Q}(u_1, u_2, \ldots, u_m)$ be a cluster algebra of geometric type of rank n. Set $I = [1, n]$ and $I' = [n+1, m]$. Fix a vertex $t_0 \in \mathbb{T}_n$ and set $\widetilde{B}(t_0) = \widetilde{B}$. We inductively define $\widetilde{B}(t)$ for all vertices t of $\mathbb{T}_n$ by setting $\widetilde{B}(t') = \mu_k(\widetilde{B}(t))$ whenever $t \overset{}{\underset{k}{\text{———}}} t'$ is an edge in $\mathbb{T}_n$. For all vertices t of $\mathbb{T}_n$ and $j \in I$, set

$$M_j(t) = \prod_{i \in I \cup I'} x_i(t)^{[b_{ij}(t)]_+}.$$

Comparing the definition of a cluster algebra of geometric type with the description of this exchange pattern cluster algebra of geometric type given in Theorem 3.3.8, we see that $\mathcal{A}(\widetilde{\mathbf{x}}, \widetilde{B})$ coincides with the exchange pattern cluster algebra associated to the $M_j(t)$, taking the ring $\mathbb{A}$ as defined above.

Conversely, given an exchange pattern cluster algebra $\mathcal{A}$ of geometric type with ring $\mathbb{A}$ as defined above, it is easy to check, again using Theorem 3.3.8, that $\mathcal{A}$ coincides with the cluster algebra of geometric type $\mathcal{A}$ associated to the seed $(\mathbf{x}(t), B(t))$ for any choice t of vertex of $\mathbb{T}_n$. □

4 Reflection groups

The classification of cluster algebras of finite type is done in terms of Cartan matrices of finite type, and the corresponding crystallographic reflection groups, or Weyl groups, and their root systems, play a key role in describing the cluster algebras. We give here a brief summary of the classification theory and some important properties of these groups and root systems. We mainly follow J. E. Humphreys' book [98].

4.1 Definition of a reflection group

Let V be a real vector space. Recall that a *bilinear form* on V is a map

$$(-,-) : V \times V \to \mathbb{R}$$

which is linear in each argument. It is *symmetric* if $(\alpha, \beta) = (\beta, \alpha)$ for all $\alpha, \beta \in V$. It is *positive definite* if $(\alpha, \alpha) > 0$ for all $\alpha \in V$. Then V is called a *Euclidean space* if it has a positive definite symmetric bilinear form on it. We assume that V is a Euclidean space. For example, we could take $V = \mathbb{R}^n$ with the usual dot product

$$((\lambda_1, \dots, \lambda_n), (\mu_1, \dots, \mu_n)) = \lambda_1\mu_1 + \dots + \lambda_n\mu_n.$$

The *length* of a vector α is given by $|\alpha| = \sqrt{(\alpha, \alpha)}$. The angle θ between two vectors α, β is given by:

$$(\alpha, \beta) = |\alpha|\,|\beta| \cos\theta,$$

and α, β are said to be *orthogonal* if they have an angle of $\pi/2$ between them. A linear map $\varphi : V \to V$ is said to be an orthogonal transformation if it preserves $(-,-)$ i.e. $(\varphi(\alpha, \beta)) = (\varphi(\alpha), \varphi(\beta))$ for all $\alpha, \beta \in V$. Then φ preserves lengths of vectors and angles between them. If φ is a multiple of such a transformation, it is called a *similarity*. We write $O(V)$ for the group of orthogonal linear transformations of V.

Lemma 4.1.1. *Let $V' \subseteq V$ be a subspace. Then $V = V' \oplus (V')^{\perp}$ where*

$$(V')^{\perp} = \{v \in V : (v, v') = 0, \text{ for all } v' \in V'\}$$

is the subspace orthogonal to V.

Definition 4.1.2. A *reflection* on V is a linear map $s : V \to V$ such that

(a) s fixes a hyperplane pointwise

(b) s reverses the direction of any normal vector to the hyperplane.

See Figure 4.1.

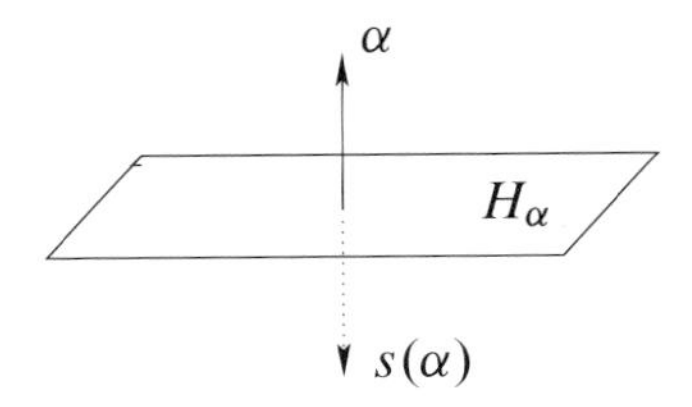

Figure 4.1. Reflection.

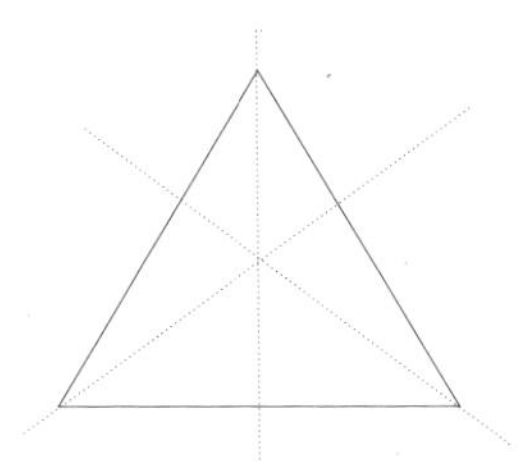

Figure 4.2. Equilateral triangle.

Here, we shall only consider hyperplanes passing through 0. Such a hyperplane has the form $H_\alpha = \{v : (\alpha, v) = 0\}$ for some vector α. We have the following (e.g. [98, §1.1]).

Lemma 4.1.3. *(a) The formula for a reflection s_α in the hyperplane H_α is given by $s_\alpha(\beta) = \beta - \frac{2(\alpha,\beta)\alpha}{(\alpha,\alpha)}$.*

(b) The reflection s_α is an orthogonal linear transformation.

Proof. For the first part, we see that if $\beta \in H_\alpha$, the formula gives $s_\alpha(\beta) = \beta$, which is correct. If $\beta = \alpha$, the formula gives $s_\alpha(\alpha) = \alpha - \frac{2(\alpha,\alpha)\alpha}{(\alpha,\alpha)} = -\alpha$, which is also correct. Since both sides are linear, the result follows.

For the second claim we notice that the orthogonality can be obtained from the formula, or using the fact that s_α is a reflection. □

A *reflection group* is a subgroup of $O(V)$ generated by reflections.

Example 4.1.4. Consider the equilateral triangle in Figure 4.2. Its symmetry group is generated by reflections in its lines of symmetry, so it is a reflection group. In fact, it is enough to take any two of the lines of symmetry.

Note that any finite dihedral group is a reflection group (it is easy to show that it is generated by reflections). We will next look at the classification of finite reflection groups.

4.2 Root systems

Suppose that W is a finite reflection group. The key to pinning W down is the notion of a *root system*, so now we will consider how to extract a root system from W. Let $\mathcal{L}_W = \{\text{span}(\alpha) \mid \alpha \in V, s_\alpha \in W\}$. Here span$(\alpha)$ denotes the span of the vector α, a subspace of V. Thus, $\mathcal{L}_W$ is the set of lines spanned by the vectors normal to the hyperplanes associated to the reflections in W.

Lemma 4.2.1. *[98, §1.2]*

(a) Let $t \in O(V)$. Then $ts_\alpha t^{-1} = s_{t\alpha}$.

(b) The set $\mathcal{L}_W$ is closed under the action of W.

Proof. For the first part we notice that if $v \in H_{t\alpha}$ then $(t\alpha, v) = 0$, so $(\alpha, t^{-1}v) = 0$. Hence

$$ts_\alpha t^{-1}(v) = ts_\alpha(t^{-1}(v)) = t(t^{-1}(v)) = v,$$

so $ts_\alpha t^{-1}$ fixes $H_{t\alpha}$ pointwise. Furthermore, $ts_\alpha t^{-1}(t\alpha) = ts_\alpha\alpha = -t\alpha$. Equality now follows from the fact that both sides of the equation are linear.

For the second part of the claim, suppose span$(\alpha) \in \mathcal{L}_W$, and $w \in W$. Then by the first part, $s_{w\alpha} = ws_\alpha w^{-1} \in W$, so span$(w\alpha) \in \mathcal{L}_W$ by definition. □

Let Φ_W be the set of unit vectors in the lines in $\mathcal{L}_W$. Note that each line gives two vectors.

Lemma 4.2.2. *[98, §1.2]*

(a) For all $\alpha \in \Phi_W$, span$(\alpha) \cap \Phi_W = \{\pm\alpha\}$.

(b) For all $\alpha, \beta \in \Phi_W$, $s_\alpha(\beta) \in \Phi_W$,

(c) The set Φ_W is finite.

Proof. (a) This follows immediately from the definition.

(b) Let $\alpha, \beta \in \Phi_W$. Then, by Lemma 4.2.1(b), span$(s_\alpha(\beta)) \in \mathcal{L}_W$. Since s_α is orthogonal, $s_\alpha(\beta)$ is a unit vector, hence in Φ_W.

(c) Since W is finite, the number of reflections in W is finite. Hence, $\mathcal{L}_W$, and thus Φ_W, is finite.

□

A subset Φ of V satisfying (a), (b) and (c) in Lemma 4.2.2 is called a *root system*. Thus we have associated a root system Φ_W to any finite reflection group, W. Note that W is generated by $\{s_\alpha : \alpha \in \Phi_W\}$. Note also that, while the elements of Φ_W are unit vectors, in general the elements of a root system needn't be unit vectors (see, for example, Section 4.8). A root system (or the corresponding reflection group) is said to be *irreducible* if it cannot be written as the union of two orthogonal subsets, each of which is a root system.

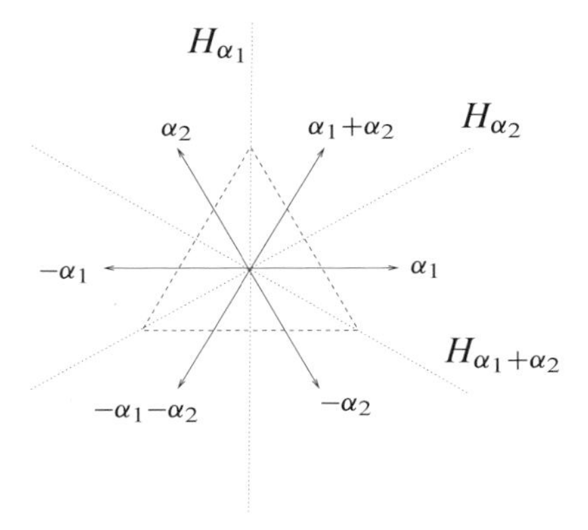

Figure 4.3. Reflection hyperplanes and normal vectors.

Example 4.2.3. In Example 4.1.4, we take the orthogonal vectors to the reflection hyperplanes (lines) corresponding to the generating reflections, and their negatives, all of the same length. It is easy to check that this set Φ of vectors forms a root system and that the corresponding reflection group is naturally the symmetry group of an equilateral triangle. See Figure 4.3, where the equilateral triangle is indicated with dashed edges. We note that

$$\Phi = \{\pm\alpha_1, \pm\alpha_2, \pm(\alpha_1 + \alpha_2)\}.$$

Conversely, if $\Phi \subseteq V$ is any root system, the group

$$W_\Phi = \langle s_\alpha \mid \alpha \in \Phi \rangle$$

is a reflection group. Then W_Φ acts on Φ via $\alpha \mapsto w(\alpha)$. This is closed because $s_\alpha(\beta) \in \Phi$ for all $\alpha, \beta \in \Phi$, and W is generated by the $s_\alpha, \alpha \in \Phi$. We thus obtain a homomorphism $\varphi : W_\Phi \to \mathrm{Sym}(\Phi)$ mapping an element of W_Φ to the permutation of Φ that it induces. Since Φ is finite, the group $\mathrm{Sym}(\Phi)$ of permutations of Φ is finite. Let $w \in W$. If $w \in \ker(\varphi)$, it fixes Φ, and therefore also its span, $\mathrm{span}(\Phi)$. If α is orthogonal to Φ, it lies in the hyperplane associated to any generating reflection of W, hence w fixes the orthogonal complement of $\mathrm{span}(\Phi)$ in V. Since $V = \mathrm{span}(\Phi) \oplus \mathrm{span}(\Phi)^\perp$, w is the identity element. Hence φ is injective, so W_Φ is finite. We have proved:

Lemma 4.2.4. *[98, §1.2] Let Φ be a root system. Then W_Φ is a finite reflection group.*

Thus, we have seen that any root system gives rise to a finite reflection group and that any finite reflection group arises in this way, since $W = W_{\Phi_W}$.

Definition 4.2.5. [148, §2.6] Let Φ, Φ' be root systems. Then we say that Φ and Φ' are *isomorphic*, written $\Phi \cong \Phi'$, if there is a similarity $V \to V'$ taking Φ to Φ'.

Lemma 4.2.6. *If $\Phi \cong \Phi'$, then $W_\Phi \cong W'_{\Phi'}$*

Proof. This is easy to show directly. □

Thus it is possible for different root systems to give rise to isomorphic reflection groups. We shall see later that it is possible for non-isomorphic root systems to give rise to isomorphic reflection groups. This may happen because roots do not have all the same length. Also note that we may not have $\operatorname{span}(\Phi) = V$, but since the orthogonal complement of $\operatorname{span}(\Phi)$ does not play a role in the study of W, we shall usually assume that $\operatorname{span}(\Phi) = V$. If not, we can always replace V with $\operatorname{span}(\Phi)$.

4.3 Simple systems

Recall that, in Example 4.1.4, we had:

$$\Phi = \{\pm\alpha_1, \pm\alpha_2, \pm(\alpha_1 + \alpha_2)\}.$$

A *simple system*, Δ, for Φ is a basis of $\operatorname{span}(\Phi)$ consisting of elements of Φ such that every root $\alpha \in \Phi$ can be written as a nonnegative linear combination of elements of Δ, or a nonpositive combination. Given a choice of Δ, its elements are referred to as *simple roots*, and the corresponding reflections as *simple reflections*. We write Φ^+ for the subset of Φ consisting of elements of the first kind (known as a *positive system*) and refer to these elements as *positive roots*. Similarly, we write Φ^- for the set of the elements of the second kind, and refer to these as *negative roots*. The *rank* of Φ is the number of elements in a simple system. In our example, $\Delta = \{\alpha_1, \alpha_2\}$ is clearly a simple system. Then we have

$$\Phi^+ = \{\alpha_1, \alpha_2, \alpha_1 + \alpha_2\}, \qquad \Phi^- = \{-\alpha_1, -\alpha_2, -\alpha_1 - \alpha_2\}.$$

But we could also take $\Delta = \{\beta_1 = \alpha_2, \beta_2 = -\alpha_1 - \alpha_2\}$. Then

$$\begin{aligned} \Phi^+ &= \{\beta_1, \beta_2, \beta_1 + \beta_2\} & \Phi^- &= \{-\beta_1, -\beta_2, -\beta_1 - \beta_2\} \\ &= \{\alpha_2, -\alpha_1 - \alpha_2, -\alpha_1\}; & &= \{-\alpha_2, \alpha_1 + \alpha_2, \alpha_1\}. \end{aligned}$$

Thus, although there is not a unique choice of simple system, the coefficients in the expressions we obtain are the same in each of these two cases. In fact, we have:

Theorem 4.3.1. *[98, §§1.3–1.8] Let Φ be a root system. Then*

(a) Φ has a simple system, Δ.

(b) If Φ^+ and Φ^- are the corresponding subsets of positive and negative roots, then $\Phi^- = -\Phi^+$.

(c) The group W_Φ acts simply transitively on the simple systems in Φ.

(d) Every root in Φ lies in the W_Φ-orbit of a simple root.

Simple systems behave well with respect to subsets. Suppose $\Phi \subseteq V$ is a root system and $\Delta \subseteq \Phi$ is a simple system for Φ, with positive roots Φ^+ and negative roots Φ^-. Let $\Delta' \subseteq \Delta$ be an arbitrary set of simple roots, and let $W_{\Delta'}$ be the subgroup of W generated by the reflections s_α for $\alpha \in \Delta'$. Let $V_{\Delta'}$ be the real subspace of V spanned by Δ' and let $\Phi_{\Delta'}$ be the intersection of Φ with $V_{\Delta'}$. Then we have the following result:

Proposition 4.3.2. *[98, §1.10] The set $\Phi_{\Delta'}$ is a root system in V with simple system Δ'. The corresponding reflection group is $W_{\Delta'}$. The same statement holds if V is replaced with $V_{\Delta'}$ (and $W_{\Delta'}$ is replaced with the restriction of $W_{\Delta'}$ to $V_{\Delta'}$).*

4.4 Coxeter groups

We fix a root system Φ with corresponding reflection group W_Φ, together with a simple system $\Delta = \{\alpha_1, \alpha_2, \dots, \alpha_n\} \subseteq \Phi$. We write s_i for s_{α_i}, for $i \in I$.

A group W is said to be a *Coxeter group* if it has a presentation of the form

$$W = \langle s_i \ : \ i \in I \mid (s_i s_j)^{m(i,j)} = e, \text{for all } i, j \in I \rangle,$$

where $m(i,i) = 1$ for all $i \in I$ and, for all $i \neq j$ in I, $m(i,j) = m(j,i)$ is either an integer which is at least 2 or $m(i,j) = \infty$ (corresponding to the absence of a relation). Note that the first condition means that $s_i^2 = e$ for all $i \in I$.

Theorem 4.4.1. *[98, §§1.9, 6.4] Let W_Φ be a finite reflection group associated to a root system Φ containing a simple system $\Delta = \{\alpha_1, \alpha_2, \dots, \alpha_n\}$. For $i, j \in I$ let $m(i,j)$ be the order of $s_i s_j$ in W. Then W is a Coxeter group with presentation*

$$W = \langle s_i : i \in I \mid (s_i s_j)^{m(i,j)} = e \text{ for all } i, j \in I \rangle. \tag{4.1}$$

Conversely, any finite Coxeter group arises in this way.

Note that, in particular, the simple reflections s_i for $i \in I$, generate W.

In our example, choosing $\Delta = \{\alpha_1, \alpha_2\}$, we see that $s_1 s_2$ is a rotation though a third of a revolution, hence of order 3. Thus

$$W_\Phi = \langle s_1, s_2 \mid s_1^2 = e, \ s_2^2 = e, \ (s_1 s_2)^3 = e \rangle$$

is a presentation of the dihedral group of order 6.

In general, an expression $w = s_{i_1} \dots s_{i_r}$ for an element of W is said to be *reduced* if it is of minimal length r; we call r the *length* of W and denote it by $l(w)$.

In our example we have $W_\Phi = \{e, s_1, s_2, s_1 s_2, s_2 s_1, s_1 s_2 s_1 = s_2 s_1 s_2\}$, with all expressions reduced.

Σ_{n+1} onto W_Φ, the reflection group of type A_n. The transpositions map to the reflections

$$(ij) \to s_{\alpha_i+\cdots+\alpha_{j-1}} = s_{e_i-e_j}, \qquad 1 \le i < j \le n+1.$$

We obtain a presentation of Σ_{n+1}:

$$\Sigma_{n+1} \cong \left\langle s_1,\dots,s_n \;\middle|\; \begin{array}{l} s_i^2 = e; \\ (s_is_j)^3 = e \text{ if } |i-j| = 1; \\ (s_is_j)^2 = e \text{ if } |i-j| > 1. \end{array} \right\rangle$$

The isomorphism takes $(i\ i+1)$ to s_i. Note that it is easy to check that the above relations hold: we see that they are in fact defining relations, by Theorem 4.4.1.

We have

$$(\alpha_i,\alpha_j) = \begin{cases} 2 & \text{if } i = j; \\ -1 & \text{if } |i-j| = 1; \\ 0 & \text{if } |i-j| > 1. \end{cases}$$

so

$$s_{\alpha_i}(\alpha_j) = \alpha_i - \frac{2(\alpha_i,\alpha_j)}{(\alpha_i,\alpha_i)}\alpha_i = \alpha_j - (\alpha_i,\alpha_j)\alpha_i = \begin{cases} -\alpha_i & \text{if } i = j; \\ \alpha_i + \alpha_j & \text{if } |i-j| = 1; \\ \alpha_j & \text{otherwise.} \end{cases}$$

Note also that $s_i(\alpha_i + \alpha_j) = \alpha_j$ if $|i-j| = 1$.

4.8 Root systems of low rank

From the classification, we see that there is a unique crystallographic reflection group of rank 1, i.e. type A_1, which is the symmetric group of order 2. It follows from Section 4.7 that it can be realised using a root system $\Phi_{A_1} = \{\alpha_1, -\alpha_1\} \subseteq V$, where V is a 1-dimensional Euclidean space spanned by a vector α_1 with $(\alpha_1,\alpha_1) = 2$.

Consider Example 4.2.3. We did not specify the lengths of the roots α_1 and α_2. If we take them to have length $\sqrt{2}$, so that $(\alpha_i,\alpha_i) = 2$ for $i = 1, 2$, then

$$(\alpha_1,\alpha_2) = |\alpha_1||\alpha_2|\cos(2\pi/3) = (\sqrt{2})(\sqrt{2})(-1/2) = -1.$$

Hence

$$\frac{2(\alpha_1,\alpha_2)}{(\alpha_1,\alpha_1)} = -1.$$

Similarly,

$$\frac{2(\alpha_2,\alpha_1)}{(\alpha_2,\alpha_2)} = -1,$$

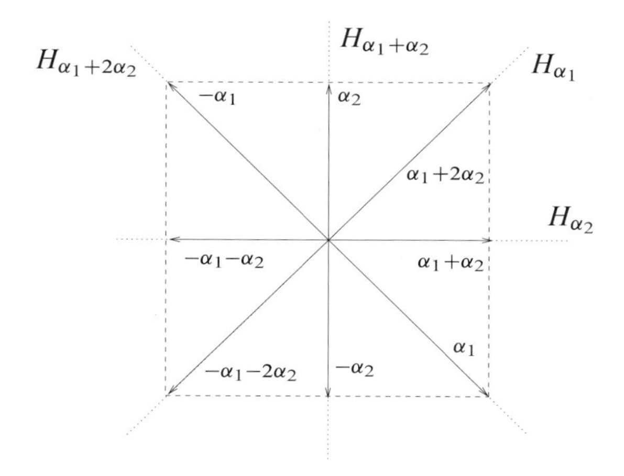

Figure 4.5. The type B_2 root system.

and it is then easy to check that the root system is crystallographic. If Φ_{A_2} is the root system described in Section 4.7, then it can be checked that the map taking α_1 to $e_1 - e_2$ and α_2 to $e_2 - e_3$ gives an isomorphism between the root system in Example 4.2.3 and Φ_{A_2}. Recall that an illustration of this root system is given in Figure 4.3.

Let $\Phi_{B_2} \subseteq \mathbb{R}^2$ be the set given by:

$$\Phi_{B_2} = \{\pm e_1, \pm e_2, \pm(e_1 - e_2), \pm(e_1 + e_2)\}.$$

Then it can be checked that Φ_{B_2} is a root system with a simple system given by $\alpha_1 = e_1 - e_2, \alpha_2 = e_2$. We have

$$\frac{2(\alpha_1, \alpha_2)}{(\alpha_1, \alpha_1)} = -1 \text{ and } \frac{2(\alpha_2, \alpha_1)}{(\alpha_2, \alpha_2)} = -2.$$

The Cartan matrix is

$$A = \begin{pmatrix} 2 & -1 \\ -2 & 2 \end{pmatrix},$$

with Dynkin diagram B_2. The corresponding reflection group is naturally identified with the symmetries of any square whose diagonals are subsets of the hyperplanes orthogonal to the roots in B_2. See Figure 4.5, where such a square is indicated with dashed edges. Note that in this case, each root actually lies in a hyperplane orthogonal to one of the other roots.

For a root system of type G_2, consider two concentric congruent equilateral triangles, one obtained from the other by flipping in a horizontal axis. The union of two triangles has boundary given by a non-convex dodecagon in the plane, and we take Φ_{G_2} to be the 12 position vectors of the vertices of the dodecagon. Then it can be checked that this is a root system of type G_2. There are two root lengths, and

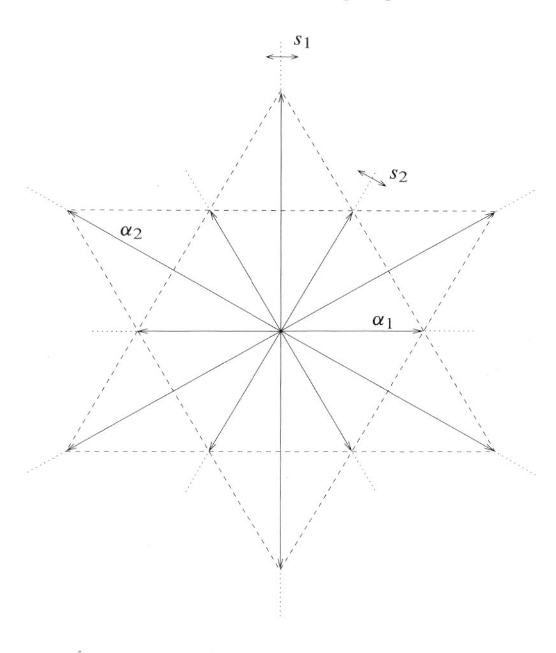

Figure 4.6. The type G_2 root system.

the corresponding reflection group can be identified with the symmetry group of the hexagon whose vertices are the short roots. See Figure 4.6.

4.9 Finite Coxeter groups

There are also some finite reflection groups which are not crystallographic. For example, the dihedral group of order $2k$, for $k \geq 3$, can be realised as the symmetry group of a regular polygon with k sides, which is generated by reflections in two lines of symmetry at an angle of π/k to each other. If $k = 3, 4$ or 6, we have seen in Section 4.8 that the corresponding reflection group is crystallographic, of type A_2, B_2 or G_2, but otherwise it is not.

We have seen (Theorem 4.4.1) that the finite reflection groups and finite Coxeter groups coincide. Given a Coxeter group with generators s_i, $i \in I$, and parameters $m(i, j)$ for $i, j \in I$, it is common to represent this data as a graph (known as the *Coxeter graph*) with vertices I and edges between every pair of vertices with $m(i, j) > 2$. If $m(i, j) = 3$, the label is usually omitted by convention. A Coxeter group is said to be *irreducible* if its Coxeter graph is connected (for finite reflection groups, this coincides with the notion of irreducibility defined in Section 4.2).

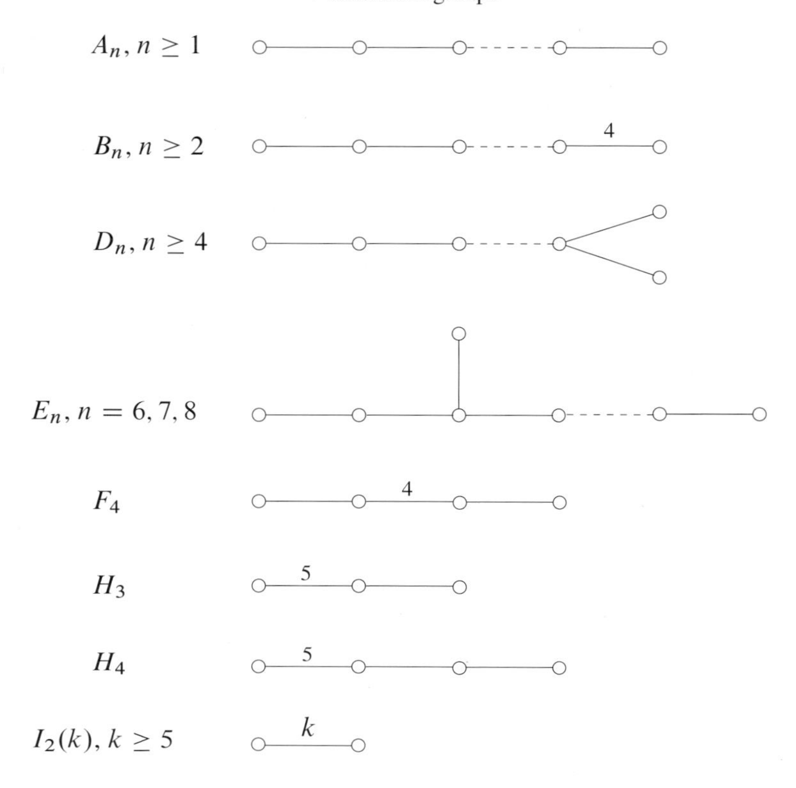

Figure 4.7. The Coxeter graphs of the finite Coxeter groups.

Theorem 4.9.1. *[98, §§2.3, 2.4, 6.4] The Coxeter graphs of the irreducible finite Coxeter groups are given in the list in Figure 4.7.*

The restriction on k for $I_2(k)$ is because $I_2(3) = A_2$ and $I_2(4) = B_2$, while $I_2(2) = A_1 \sqcup A_1$ is not irreducible. Type G_2 appears as $I_2(6)$. Note that the finite reflection groups of types B_n and C_n are isomorphic, so have only one Coxeter graph in the list (denoted B_n). Thus, the finite irreducible non-crystallographic reflection groups are H_3, H_4 and $I_2(k)$ for $k = 5$ or $k \geq 7$. The other graphs in Figure 4.7 correspond to the finite irreducible crystallographic reflection groups.

4.10 Reduced expressions

Let $W = W_\Phi$ be a reflection group, with root system Φ, and $\Delta = \{\alpha_1, \dots, \alpha_n\} \subseteq \Phi$ a simple system. So W is generated by the $s_{\alpha_i} = s_i$ for $i = 1, 2, \dots, n$. Recall that an expression $w = s_{i_1} \dots s_{i_r}$ for w is *reduced* if r is minimal; then r is said to be the *length* of w.

Proposition 4.10.1. *[98, §1.7] Let $w = s_{i_1} \dots s_{i_r}$ be a reduced expression for w. Then the positive roots made negative by w are exactly the β_t, $t = 1, \dots, r$, where $\beta_t = s_{i_r} \dots s_{i_{t+1}}(\alpha_{i_t})$. Thus there are exactly $l(w)$ such roots.*

Since $\Phi = \Phi^+ \cup \Phi^-$, the largest value $l(w)$ can take is $|\Phi^+|$. In fact, there is always a unique element, w_0, of this length, making all positive roots negative. It is known as the *longest element* of W.

Example 4.10.2. (Type A_3). Let $w = s_2 s_1 s_3$. Then the positive roots made negative by w are α_3, $s_3(\alpha_1) = \alpha_1$ and

$$s_3 s_1(\alpha_2) = s_3(\alpha_1 + \alpha_2) = \alpha_1 + \alpha_2 + \alpha_3.$$

In this case a reduced expression for w_0 is given by:

$$w_0 = s_2 s_1 s_3 s_2 s_1 s_3.$$

Notice that its length is $6 = |\Phi^+|$. We have $w_0 = \begin{pmatrix} 1 & 2 & 3 & 4 \\ 4 & 3 & 2 & 1 \end{pmatrix}$ as a permutation. In type A_n,

$$w_0 = \begin{pmatrix} 1 & 2 & \dots & n+1 \\ n+1 & n & \dots & 1 \end{pmatrix}$$

and has length equal to $\frac{1}{2}n(n+1)$.

In fact, we have the following result [49, 114, 156]. Recall that a *partition* of n is a tuple

$$\lambda = (\lambda_1, \lambda_2, \dots, \lambda_r)$$

where $\sum_{i=1}^r \lambda_i = n$ and $\lambda_1 \geq \lambda_2 \geq \dots \geq \lambda_r > 0$. Such partitions are often displayed as *Young diagrams*. In the Young diagram corresponding to λ, there are λ_1 boxes in the first row, λ_2 boxes in the second row, and so on, each row immediately below the previous one. For an example, corresponding to the partition $(5, 4, 3, 2, 1)$, see Figure 4.8. A *hook* in a Young diagram consists of one of the boxes together with all of the boxes below it and to its right. The *length* of the hook is the number of boxes in it.

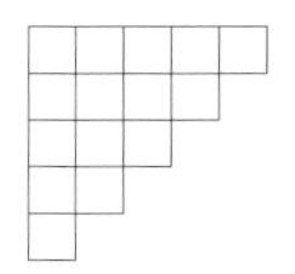

Figure 4.8. The Young diagram corresponding to the partition $(5, 4, 3, 2, 1)$ of 15.

Theorem 4.10.3. *[49, 114, 156] The number of reduced expressions for w_0 in type A_n is given by*

$$\frac{(\frac{1}{2}n(n+1))!}{r}$$

where r is the product of the lengths of the hooks in the Young diagram corresponding to the partition $(n, n-1, n-2, \ldots, 2, 1)$ of $\frac{1}{2}n(n+1)$.

Example 4.10.4. In type A_2, the Young diagram:

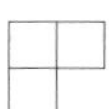

has three hooks, two of length 1 and one of length 3. Hence the number of reduced expressions for w_0 is given by $\frac{3!}{1.1.3} = 2$. The two reduced expressions for w_0 are $s_1s_2s_1$ and $s_2s_1s_2$.

In type A_3, the number of reduced expressions for w_0 is given by $\frac{6!}{1.1.1.3.3.5} = 16$ (an example of such a reduced expression is $s_2s_1s_3s_2s_1s_3$). In type A_4, there are 768 reduced expressions for w_0.

The presentation for a finite reflection group as a Coxeter group (equation (4.1)) can be rewritten:

$$W = \left\langle s_i \ : i \in I \ \middle| \ \begin{array}{c} s_i^2 = e, \text{ for all } i \in I \\ s_is_j \cdots = s_js_i \cdots, \text{ for all } i \neq j \in I \end{array} \right\rangle, \tag{4.2}$$

where, in the product in the second line, there are $m(i, j)$ terms on each side of the relation. The relations in this second line are sometimes referred to as the *braid relations*. The corresponding Artin braid group is defined using only these relations, and thus has presentation:

$$B = \langle \sigma_i \ : i \in I \mid \sigma_i\sigma_j \cdots = \sigma_j\sigma_i \cdots, \text{ for all } i \neq j \in I \rangle,$$

where again there are $m(i, j)$ terms on each side of the relation. The Artin braid group is thus a quotient of the corresponding reflection group.

In type A_n, this gives a presentation of the usual braid group on $n+1$ strings:

$$B = \left\langle \sigma_1 \ldots \sigma_{n+1} \ \middle| \ \begin{array}{c} \sigma_i\sigma_j\sigma_i = \sigma_j\sigma_i\sigma_j, \ |i-j| = 1 \\ \sigma_i\sigma_j = \sigma_j\sigma_i, \ |i-j| > 1 \end{array} \right\rangle$$

and the symmetric group of degree Σ_{n+1} is a quotient via the map $\sigma_i \mapsto s_i$.

Theorem 4.10.5. *[127, 160] Let W be a reflection group. Let $w = s_{i_1} \ldots s_{i_r} = s_{j_1} \ldots s_{j_r}$ be reduced expressions for an element $w \in W$. Then, there is a sequence of applications of braid relations taking the first expression to the second.*

In the sequence of relations given by Theorem 4.10.5, we never use the relation $s_i^2 = e$. Note also that applying a braid relation to a reduced expression always gives another reduced expression.

The relations

$$s_i s_j = s_j s_i$$

when $m(i, j) = 2$, appearing in the presentation (4.2) are known as *commutation relations*.

Two reduced expressions for an element w are said to be *commutation equivalent* if there is a sequence of commutations taking the first to the second. In type A_n they are of the form $\sigma_i \sigma_j = \sigma_j \sigma_i$ for $|i - j| > 1$. The equivalence classes are known as *commutation classes*.

Open problem: Find the number of commutation classes of reduced expressions for w_0 in type A_n.

The numbers of reduced expressions and commutation classes of reduced expressions for w_0 in type A_n for small n are given in the following table.

n	1	2	3	4	5
Number of commutation classes	1	2	8	62	908
Number of reduced expressions	1	2	16	768	292864

Note that the number of commutation classes for w_0 appears in sequence A006245 in [140].

4.11 Coxeter elements

Let Φ be a root system and $W = W_\Phi$ the corresponding reflection group. A *Coxeter element* in W is an element of the form $c_\Delta = s_1 \dots s_n$, where $\Delta = \{\alpha_1, \alpha_2, \dots, \alpha_n\}$ is a simple system in Φ. Note that the element we obtain is dependent on a choice of the simple system as well as an ordering. Coxeter elements correspond (non-bijectively) to orientations of the Dynkin diagram (see Section 5.5). We have:

Proposition 4.11.1. *[98, §3.16] Any two Coxeter elements in W are conjugate in W.*

The order of one (and hence any) Coxeter element is known as the *Coxeter number h* of Φ (or W). For example, in type A_n with the choice of simple system, Δ, as in Section 4.7, we have:

$$\begin{aligned} c_\Delta &= (1\ 2)(2\ 3)\dots(n-1\ n)(n\ n+1) \\ &= (1\ 2\ \dots\ n+1). \end{aligned}$$

Hence, the Coxeter number in type A_n is equal to $n+1$. Since all Coxeter elements in W are conjugate, they all have the same eigenvalues, which must be of the form ω^e, where ω is a primitive h^{th} root of unity and $0 \le e < h$. The *exponents* of W are the various exponents e that appear (with multiplicity), in these expressions for the eigenvalues of a Coxeter element. We denote them by $e_1, \dots, e_n$. The *degrees* of W are given by $d_i = e_i + 1, i = 1, \dots n$.

Example 4.11.2. The exponents in type A_n are $1, 2, \dots, n$. Thus the degrees are $2, 3, \dots, n+1$.

Proposition 4.11.3. *[98]*

(a) We have $\sum_{i=1}^{k} e_i = nh/2 = |\Phi^+|$, the number of positive roots;

(b) The numbers $1, h-1$ are always exponents;

(c) We have $|W| = \prod_{i=1}^{k} d_i$;

(d) If W is irreducible and crystallographic and e (with $1 \leq e \leq h-1$) is coprime to h, then e is an exponent of W.

Proof. For (a), see [98, §§3.9, 3.18], for (b) see [98, §3.17], for (c) see [98, §3.9] and for (d) see [98, §3.20]. □

For example, for type A_n, we see in Proposition 4.11.3(c) that $|W| = |\Sigma_{n+1}| = 2 \cdot 3 \cdots (n+1) = (n+1)!$.

The *tensor algebra* of a vector space X is

$$T(X) = \mathbb{R} \oplus X \oplus X \otimes X \oplus \dots$$

where the multiplication is just concatenation of tensors, expanded linearly, i.e.

$$(x_1 \otimes \dots \otimes x_r)(x_1' \otimes \dots \otimes x_r') = x_1 \otimes \dots \otimes x_r \otimes x_1' \otimes \dots \otimes x_r'.$$

The *symmetric algebra* $S(X)$ of X is the quotient $T(X)/I$, where I is the ideal generated by $xx' - x'x$ for all $x, x' \in X$.

Suppose that $e_1, e_2, \dots, e_n$ is a basis of V. Let $x_1, x_2, \dots, x_n$ be the dual basis of V^*. Then it is easy to see that $S(V^*) = \mathbb{R}[x_1, \dots, x_n]$ is a polynomial ring.

We assume $\Phi \subseteq V$ is a root system whose span is V. Then $W = W_\Phi$ acts on V^* via $(wf)(v) = f(w^{-1}v)$, and this induces an action of W on $S(V^*)$.

Theorem 4.11.4 (C. Chevalley). *[98, Chap. 3] The subalgebra of $S(V^*)$ consisting of the W-invariant polynomials is generated as an $\mathbb{R}$-algebra (with identity) by n homogeneous algebraically independent polynomials of positive degree. The degrees of these polynomials are uniquely determined and coincide with the degrees d_i of W defined above.*

5 Cluster algebras of finite type

A cluster algebra is said to be of *finite type* if it has finitely many seeds. It is said to be of *finite mutation type* if the set of principal parts of its exchange matrices is finite. These are different notions (the former implying the latter). In this chapter we consider the classification of the cluster algebras of finite type by S. Fomin and A. Zelevinsky [70] in terms of finite type Cartan matrices (or Dynkin diagrams). Cluster algebras of finite mutation type (in the skew-symmetric case) will be considered later, in Chapter 8.

5.1 Classification

Recall that a cluster algebra is defined by choosing a field $\mathbb{F} = \mathbb{Q}(u_1, \ldots, u_m)$, a free generating set (an extended cluster) $\widetilde{\mathbf{x}} = \{x_1, \ldots x_m\}$, $n \leq m$, and an $m \times n$ integer matrix $\widetilde{B}$ whose principal part, $B = (b_{ij})_{i,j=1}^n$ is sign-skew-symmetric. Furthermore, every iterated mutation of $\widetilde{B}$ must have sign-skew-symmetric principal part. The cluster algebra corresponding to this data is denoted by $\mathcal{A}(\widetilde{\mathbf{x}}, \widetilde{B})$. It does not depend on the choice of $(\widetilde{\mathbf{x}}, \widetilde{B})$ up to strong isomorphism, so we may denote it by $\mathcal{A}(\widetilde{B})$ when an explicit choice of free generating set is not important.

Suppose that B is an $n \times n$ sign-skew-symmetric integer matrix whose mutation class contains only sign-skew-symmetric matrices. Then, any $m \times n$ integer matrix $\widetilde{B}$ with principal part B defines a cluster algebra. We define the family $Cl(B)$ to be the collection of all such cluster algebras. Two families $Cl(B)$ and $Cl(B')$ are said to be *strongly isomorphic* provided each cluster algebra $\mathcal{A}(\widetilde{B})$ in $Cl(B)$ is strongly isomorphic to a cluster algebra $\mathcal{A}(\widetilde{B'})$ in $Cl(B')$ and vice versa.

For two $m \times n$ matrices $\widetilde{B}$ and $\widetilde{B}'$, we write $\widetilde{B} \sim_{mut} \widetilde{B}'$ to indicate that $\widetilde{B}$ and $\widetilde{B}'$ lie in the same mutation class, i.e. that there is a finite sequence of mutations taking $\widetilde{B}$ to $\widetilde{B}'$.

Lemma 5.1.1. *[70, §1] Two families $Cl(B)$ and $Cl(B')$ are strongly isomorphic if and only if $B \sim_{mut} B'$.*

Proof. Assume first that $Cl(B) \cong Cl(B')$. Since strong isomorphism preserves the number of elements in a free generating set in a seed, we have $\mathcal{A}(B) \cong \mathcal{A}(B')$, so $B \sim_{mut} B'$.

Conversely, suppose $B \sim_{mut} B'$, and let $\mathcal{A}(\widetilde{B}) \in Cl(B)$. Then there is a sequence of mutations taking B to B'. We apply the same sequence to $\widetilde{B}$, and obtain a matrix $\widetilde{B}'$ with principal part B'. Since this is the matrix in a seed of $\mathcal{A}(\widetilde{B})$, we have $\mathcal{A}(\widetilde{B}) \cong \mathcal{A}(\widetilde{B}')$. Similarly, every cluster algebra in $Cl(B')$ is strongly isomorphic to one in $Cl(B)$. □

Recall that a cluster algebra $\mathcal{A}(B)$ is said to be of *finite type* if its set of seeds is finite.

If B is an $n \times n$ integer square matrix, its *Cartan counterpart* A is the $n \times n$ integer matrix $A = A(B) = (a_{ij})_{1 \le i,j \le n}$ where

$$a_{ij} = \begin{cases} 2 & \text{if } i = j; \\ -|b_{ij}| & \text{if } i \neq j. \end{cases}$$

Then, for $i, j \in I$, we have:

$$\begin{aligned} a_{ii} &= 2; \\ a_{ij} &\le 0, i \neq j; \\ a_{ij} &= 0 \text{ if and only if } a_{ji} = 0. \end{aligned}$$

Note that if B is skew-symmetrizable, there is a diagonal matrix

$$D = \begin{pmatrix} d_1 & & & \\ & d_2 & & \\ & & \ddots & \\ & & & d_n \end{pmatrix}$$

with positive integer diagonal entries such that DB is skew-symmetric, i.e.:

$$d_i b_{ij} = -d_j b_{ji} \text{ for all } i, j.$$

Then $d_i a_{ij} = d_j a_{ji}$ for all i, j (as d_i, a_{ij} are both nonnegative), i.e. DA is symmetric, so B skew-symmetrizable implies $A(B)$ is symmetrizable.

Theorem 5.1.2. *[70, Thm. 1.4]*

(a) *All cluster algebras in a family $Cl(B)$ are simultaneously of finite or infinite type.*

(b) *A cluster algebra is of finite type if and only if the Cartan counterpart of the principal part of one of its seeds is a Cartan matrix of finite type.*

(c) *The families of cluster algebras of finite type are classified up to strong isomorphism by the Cartan matrices of finite type (up to simultaneous permutation of rows and columns), with the Cartan matrix associated to a given family of cluster algebras arising as in (b) from any cluster algebra in the family.*

We note that Theorem 5.1.2(b) is quite a strong result: whether a cluster algebra is of finite type depends only on the principal part of the matrices in its seeds, and is independent of the coefficients. In fact, a more general result is true: the exchange graph of a cluster algebra of geometric type defined by a skew-symmetric matrix depends only on the principal part of the matrix [34, Thm. 4.6] (in fact this is extended in [34, Cor. 5.5] to cluster algebras with a more general notion of coefficients, as defined in [72, §2]). This result was conjectured (in general) in [69, Conj.

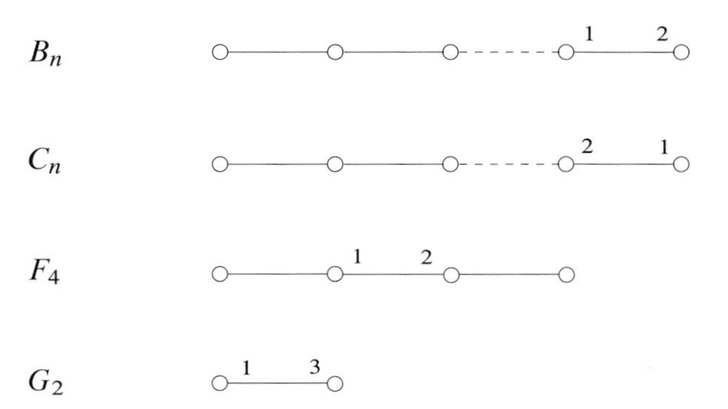

Figure 5.1. The valued graphs corresponding to the non-simply-laced Dynkin diagrams.

4.14(1)]. It is known in the skew-symmetrizable case, provided the principal part is non-degenerate [85, Thm. 4].

A Cartan matrix $A = (a_{ij})$ defines, in an natural way, a valued graph (see Section 2.4) with set of vertices I. Whenever $a_{ij} \neq 0$, there is an edge between i and j and we set $v_{ij} = -a_{ij}$. A Dynkin diagram is said to be *simply-laced* if it has no multiple arrows, i.e. it is A_n, D_n, E_6, E_7, or E_8. In this case, the valued graph structure of the corresponding Cartan matrix is trivial. In Figure 5.1, we show the valued graphs corresponding to the Cartan matrices of the non-simply-laced Dynkin diagrams.

In the sequel we shall identify a Dynkin diagram with the valued graph associated to the corresponding Cartan matrix in this way. We may then restate the above theorem in the language of valued quivers. Recall that a valued quiver is an orientation of a valued graph.

Theorem 5.1.3. *[70, Thm. 1.4]*

(a) *All cluster algebras in a family $Cl(B)$ are simultaneously of finite or infinite type.*

(b) *A cluster algebra is of finite type if and only if the underlying valued graph of the principal part of one of its valued quiver seeds is a Dynkin diagram (with the above identification).*

(c) *The families of cluster algebras of finite type are classified up to strong isomorphism by the Dynkin diagrams, with the Dynkin diagram associated to a given family of cluster algebras arising as in (b) from any cluster algebra in the family.*

Note that in the case where the Dynkin diagram is simply-laced, the underlying valued graph will just be an underlying graph.

We write Cl_Γ for the family of cluster algebras corresponding to a Dynkin diagram Γ. If a cluster algebra $\mathcal{A}$ lies in Cl_Γ for a Dynkin diagram Γ, we shall say that it has *type* Γ. Note that in (b), only one Dynkin diagram can ever arise from a seed in $\mathcal{A}$: in other words, all orientations of a given Dynkin diagram (as valued quiver), are mutationally equivalent. This follows from the following well-known fact:

Lemma 5.1.4. *All orientations of a (finite) tree are mutationally equivalent via a sequence of mutations at sinks or sources only.*

Proof. We prove this by induction on the number of vertices, n. It is clearly true for one vertex, so suppose it is true for fewer than n vertices and let T be a tree with n vertices, with two orientations Q and Q'. Number the vertices of T so that there is an edge joining vertices $n-1$ and n, with n a vertex of valency 1. We may assume that n is a sink in both Q and Q'. By induction, the two quivers obtained from Q and Q' by removing vertex n are mutationally equivalent. Applying the same sequence of mutations, we can go from Q to Q', mutating at n when necessary to ensure that $n-1$ is always a sink or source. □

Since each family $Cl(B)$ contains a unique strong isomorphism class $A(B)$ of cluster algebras without coefficients, Theorems 5.1.2, 5.1.3 induce a classification of cluster algebras without coefficients up to strong isomorphism. We shall denote by $\mathcal{A}_\Gamma$ the cluster algebra without coefficients corresponding to a Dynkin diagram Γ. This can be constructed by taking A to be the corresponding Cartan matrix of finite type (well defined up to simultaneous permutation of rows and columns), and choosing an arbitrary sign-skew-symmetric matrix B such that $A(B) = A$ (note that B is necessarily skew-symmetrizable). Then $\mathcal{A}_\Gamma$ can be taken to be $\mathcal{A}(B)$. Note that the results in this section also apply to cluster algebras over any field of characteristic zero.

Example 5.1.5. In Example 2.1.7 we considered the exchange matrix $B = \begin{pmatrix} 0 & 1 \\ -1 & 0 \end{pmatrix}$. Its Cartan counterpart is $A(B) = \begin{pmatrix} 2 & -1 \\ -1 & 2 \end{pmatrix}$, which we have seen is the Cartan matrix of type A_2. Thus $Cl_{A_2} = Cl(B)$ with B as above. There is one other choice of sign for a sign-skew-symmetric matrix B with $A(B) = A$, i.e. $\begin{pmatrix} 0 & -1 \\ 1 & 0 \end{pmatrix}$, and we see that this matrix also appears in a seed of this cluster algebra.

Alternatively, the quiver corresponding to the above matrix B is $1 \to 2$, which has underlying graph of type A_2.

The matrices $\begin{pmatrix} 0 & 1 \\ -2 & 0 \end{pmatrix}$ and $\begin{pmatrix} 0 & 1 \\ -3 & 0 \end{pmatrix}$ have Cartan counterparts of type B_2 and G_2, respectively. The corresponding valued quivers are shown in Figure 5.2. Thus the three recurrences considered in Section 1.2 correspond to the three irreducible root systems of rank 2; see also Example 2.1.7.

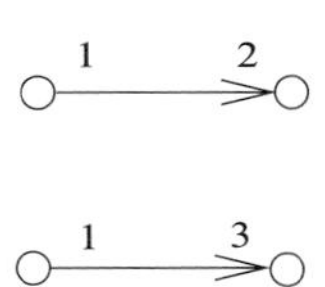

Figure 5.2. Valued quivers in Example 5.1.5.

Example 5.1.6. In Example 2.5.2, the exchange matrix $\begin{pmatrix} 0 & 1 \\ -1 & 0 \\ 1 & 0 \\ 0 & 1 \end{pmatrix}$, has principal part equal to the matrix B in Example 5.1.5 and thus the corresponding cluster algebra also lies in Cl_{A_2}.

Example 5.1.7. Example 2.1.8 has exchange matrix $\begin{pmatrix} 0 \\ 1 \\ 1 \end{pmatrix}$, whose principal part has Cartan counterpart (2), i.e. the Cartan matrix of type A_1. Hence this cluster algebra lies in Cl_{A_1}.

We remark that the articles [12] and [154] give ways of checking directly if a cluster algebra has finite type from a single seed. The article [13] gives presentations of crystallographic reflection groups corresponding to arbitrary seeds in the corresponding cluster algebra; if the Cartan counterpart is a finite type Cartan matrix then the presentation reduces to the usual Coxeter presentation; see Section 4.4.

5.2 Folding

We follow [45], which also uses [117], where admissible group actions are considered. Let B be a skew-symmetrizable matrix. A permutation of I is said to be an *automorphism* of B provided $b_{g^{-1}i,g^{-1}j} = b_{ij}$ for all $i, j \in I$. Equivalently, it must be an automorphism of the corresponding valued quiver, Q (thus, preserving the labels on the arrows). A group G acting on B via automorphisms is said to be acting *admissibly* if, for every i, j in a the same G-orbit $b_{ij} = 0$, i.e. there is no arrow between i and j.

Suppose G acts admissibly on B. Let $\overline{I}$ denote the set of G orbits of G on I, writing $\overline{i}$ for the G-orbit containing $i \in I$. The quotient (or folded) matrix $\overline{B}$ is the square matrix with rows and columns indexed by $\overline{I}$ and entries defined by:

$$\overline{b}_{\overline{i},\overline{j}} = \sum_{k \in \overline{i}} b_{k,j}.$$

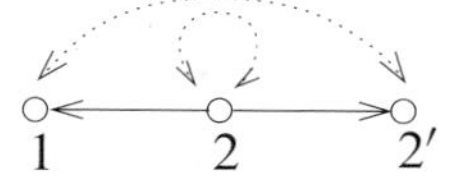

$$\begin{pmatrix} 0 & 1 & 1 \\ -1 & 0 & 0 \\ -1 & 0 & 0 \end{pmatrix}$$

Figure 5.3. A quiver of type A_3, to be folded.

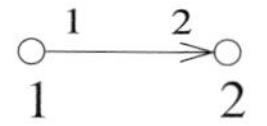

$$\begin{pmatrix} 0 & 1 \\ -2 & 0 \end{pmatrix}$$

Figure 5.4. The folding of the quiver in Figure 5.3.

We write $\overline{Q}$ for the corresponding valued quiver.

Lemma 5.2.1. *[45, §2] Suppose G acts on B admissibly. Then $\overline{B}$ is a well-defined skew-symmetrizable matrix.*

By [45, Lemma 3.4], the b_{kj} in the above sum are either all nonnegative or all nonpositive. If they are all nonnegative (and at least one is positive), there is an arrow from $\overline{i}$ to $\overline{j}$ in $\overline{Q}$ with labels $\sum_{k\in\overline{i}} b_{k,j}$ and $-\sum_{k\in\overline{j}} b_{k,i}$ (with a similar statement in the nonpositive case).

As an example, we consider the quiver and skew-symmetric matrix in Figure 5.3 (with the ordering $1, 2, 2'$ on the rows and columns). If we take G to be the automorphism group interchanging vertices 2 and $2'$, the folded quiver is as shown in Figure 5.4. Thus we see that folding an orientation of the Dynkin diagram of type A_3 gives an orientation of the Dynkin diagram of type B_2 (as a valued quiver).

Orientations of B_n can be obtained from D_{n+1} (taking D_3 to be A_3), C_n from A_{2n-1}, E_6 from F_4 and G_2 from D_4 in a similar way.

Suppose G acts admissibly on B. Let $x_i, i \in I$ and $y_{\overline{i}}, i \in \overline{I}$ be indeterminates, so that $\mathcal{A}(B)$ is a subring of the field $\mathbb{F} = \mathbb{Q}(u_i, i \in I)$ while $\mathcal{A}(B/G)$ is a subring of the field $\overline{F} = \mathbb{Q}(v_{\overline{i}}, \overline{i} \in \overline{I})$. There is a natural projection from $\mathbb{F}$ to $\overline{\mathbb{F}}$ taking u_i to $v_{\overline{i}}$ for $i \in I$.

Theorem 5.2.2. *[45, Thm. 7.3] Suppose G acts admissibly on B, where B corresponds to a simply-laced Dynkin diagram. Then $\pi(\mathcal{A}(B)) = \mathcal{A}(B/G)$.*

Thus we see that folding gives a way of constructing the cluster algebras corresponding to the non-simply-laced Dynkin diagrams in terms of the cluster algebras of the simply-laced Dynkin diagrams.

We note that results using folding-type arguments to study non-simply-laced cluster algebras have also been obtained in [38, 39, 164]. We also note that care needs to be taken with folding: the counter-example in [37, §12] can be regarded as a valued quiver which cannot be obtained by folding in a certain sense. See also [52, 53].

5.3 Denominators

Next, we would like to describe the cluster algebras of finite type in greater detail. Recall that the cluster variables in type A_2 (with initial cluster $\{x_1, x_2\}$) are:

$$x_1,\, x_2,\, \frac{x_2+1}{x_1},\, \frac{x_1+1}{x_2},\ \text{and}\ \frac{x_1+x_2+1}{x_1x_2}.$$

The root system of type A_2 is $\{\pm\alpha_1, \pm\alpha_2, \pm(\alpha_1+\alpha_2)\}$. Let

$$\Phi_{\geq -1} = \{-\alpha_1, -\alpha_2, \alpha_1, \alpha_2, \alpha_1+\alpha_2\}.$$

Then there is a nice correspondence between $\Phi_{\geq -1}$ and the cluster variables:

Root	Cluster variable
$-\alpha_1$	$\frac{1}{x_1^{-1}} = x_1$
$-\alpha_2$	$\frac{1}{x_2^{-1}} = x_2$
α_1	$\frac{x_2+1}{x_1}$
α_2	$\frac{x_1+1}{x_2}$
$\alpha_1+\alpha_2$	$\frac{x_1+x_2+1}{x_1x_2}$

This is a special case of the following result (we give citation information in the remark below).

Theorem 5.3.1. *Let $\mathcal{A}$ be a cluster algebra of finite type Γ, without coefficients. Let $(\{x_1, \dots x_n\}, Q)$ be an initial cluster for $\mathcal{A}$ in which Q is a valued quiver which is an orientation of the valued graph Γ. Let Φ be the root system corresponding to Γ, containing positive roots Φ^+ and simple system $\Delta = \{\alpha_1, \dots, \alpha_n\}$, with almost positive roots $\Phi_{\geq -1} = \Phi^+ \cap (-\Delta)$. For a root $\alpha = d_1\alpha_1 + \cdots + d_n\alpha_n \in \Phi$, let $x^\alpha = x_1^{d_1} + \cdots + x_n^{d_n}$. Then there is a bijection $\alpha \mapsto x[\alpha]$ between $\Phi_{\geq -1}$ and the cluster variables of $\mathcal{A}$, such that*

$$x[\alpha] = \frac{P_\alpha(x_1, \dots, x_n)}{x^\alpha},$$

where P_α is a polynomial in $x_1, \dots, x_n$ with positive integer coefficients, and non-zero constant term. In particular $x[-\alpha_i] = x_i$ for $i = 1, \dots, n$.

Remark 5.3.2. (a) For Q an *alternating quiver*, i.e. with every vertex a sink or a source, this was first proved in [70, Thm. 1.9]. In this case, coefficients are allowed and $P_\alpha \in (\mathbb{Z}[x_{n+1}, \dots, x_m])[x_1, \dots, x_n]$ has non-zero constant term. See also [27, Thm. 6.6] and [28, Thm. 4.4].

(b) In the generality stated here, Theorem 5.3.1 was proved in [45, §§6-7] and [167, Prop. 4.7].

(c) More general results in this direction hold. E.g. for acyclic quivers, see [36, Thms. 3,4], [24, Thms. 2.2-2.3] and the Appendix of [24].

Example 5.3.3. A cluster algebra of type A_3. The root system of type A_3 can be given as $\Phi = \Phi^+ \cup \Phi^-$, where $\Phi^+ = \{\alpha_1, \alpha_2, \alpha_3, \alpha_1 + \alpha_2, \alpha_2 + \alpha_3, \alpha_1 + \alpha_2 + \alpha_3\}$ and $\Delta = \{\alpha_1, \alpha_2, \alpha_3\}$ is a simple system. The Cartan matrix is

$$\begin{pmatrix} 2 & -1 & 0 \\ -1 & 2 & -1 \\ 0 & -1 & 2 \end{pmatrix},$$

so, by Theorem 5.1.2, we can take the initial seed

$$\left(\{x_1, x_2, x_3\}, \begin{pmatrix} 0 & 1 & 0 \\ -1 & 0 & -1 \\ 0 & 1 & 0 \end{pmatrix}\right).$$

In fact we can take instead the corresponding quiver seed:

$$(\{x_1, x_2, x_3\},\ 1 \longrightarrow 2 \longleftarrow 3\ ).$$

The cluster variables are:

$$\begin{aligned}
&x_{-\alpha_1} = x_1; && x_{\alpha_1} = \tfrac{x_2+1}{x_1}; && x_{\alpha_1+\alpha_2} = \tfrac{x_1x_3+x_2+1}{x_1x_2};\\
&x_{-\alpha_2} = x_2; && x_{\alpha_2} = \tfrac{x_1x_3+1}{x_2}; && x_{\alpha_2+\alpha_3} = \tfrac{x_1x_3+x_2+1}{x_2x_3};\\
&x_{-\alpha_3} = x_3; && x_{\alpha_3} = \tfrac{x_2+1}{x_3}; && x_{\alpha_1+\alpha_2+\alpha_3} = \tfrac{x_1x_3+(x_2+1)^2}{x_1x_2x_3}.
\end{aligned}$$

The clusters are determined by certain subsets of $\Phi_{\geq -1}$ of cardinality 3, and there are 14 clusters in total. These subsets are listed below (on the left), sorted out according to the corresponding matrix (or quiver) in the seed, displayed on the right.

Clusters	Exchange matrix	Quiver
$\{-\alpha_1, -\alpha_2, -\alpha_3\}$ $\{\alpha_1 + \alpha_2, \alpha_2, \alpha_2 + \alpha_3\}$ $\{\alpha_3, \alpha_1 + \alpha_2 + \alpha_3, \alpha_1\}$	$\begin{pmatrix} 0 & 1 & 0 \\ -1 & 0 & -1 \\ 0 & 1 & 0 \end{pmatrix}$	$1 \longrightarrow 2 \longleftarrow 3$
$\{-\alpha_1, \alpha_2, -\alpha_3\}$ $\{\alpha_1 + \alpha_2, \alpha_1 + \alpha_2 + \alpha_3, \alpha_2 + \alpha_3\}$ $\{\alpha_3, -\alpha_2, \alpha_1\}$	$\begin{pmatrix} 0 & -1 & 0 \\ 1 & 0 & 1 \\ 0 & -1 & 0 \end{pmatrix}$	$1 \longleftarrow 2 \longrightarrow 3$
$\{-\alpha_3, \alpha_2, \alpha_1 + \alpha_2\}$ $\{\alpha_2 + \alpha_3, \alpha_1 + \alpha_2 + \alpha_3, \alpha_3\}$ $\{\alpha_1, -\alpha_2, -\alpha_3\}$ $\{-\alpha_1, \alpha_2, \alpha_2 + \alpha_3\}$ $\{\alpha_1 + \alpha_2, \alpha_1 + \alpha_2 + \alpha_3, \alpha_1\}$ $\{\alpha_3, -\alpha_2, -\alpha_1\}$	$\begin{pmatrix} 0 & -1 & 0 \\ 1 & 0 & -1 \\ 0 & 1 & 0 \end{pmatrix}$	$1 \longleftarrow 2 \longleftarrow 3$
$\{\alpha_1, \alpha_1 + \alpha_2, -\alpha_3\}$ $\{\alpha_3, \alpha_2 + \alpha_3, -\alpha_1\}$	$\begin{pmatrix} 0 & 1 & -1 \\ -1 & 0 & 1 \\ 1 & -1 & 0 \end{pmatrix}$	$1 \longrightarrow 2 \longrightarrow 3$, $3 \longrightarrow 1$

Note that every possible orientation of the Dynkin diagram

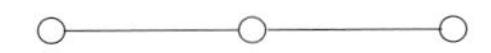

of type A_3 arises as the quiver of a seed in this cluster algebra, as predicted by Theorem 5.1.3. The article [149] gives some explicit formulas for cluster variables in type A_n.

An interesting question then is to determine which collections of roots (known as *Q-root clusters*) give rise to clusters — we shall consider this in the next section. We first note an interesting result of S. Fomin and A. Zelevinsky in this context.

Theorem 5.3.4. *[70, Thm. 1.12] Let $\mathcal{A}$ be a cluster algebra of finite type. Then every seed in $\mathcal{A}$ is uniquely determined by its cluster,* $\mathbf{x}$. *Furthermore, if $\mathbf{x}$ is a cluster and $x \in \mathbf{x}$, there is a unique cluster $\mathbf{x}'$ with $\mathbf{x} \cap \mathbf{x}' = \mathbf{x}\backslash\{x\}$.*

A skew-symmetric cluster algebra is said to be *acyclic* if it contains a seed whose underlying quiver has principal part which is acyclic, i.e. contains no oriented cycles. Because Dynkin diagrams are acyclic, any cluster algebra of finite type is acyclic. The above result holds for acyclic cluster algebras: see [36, Cor. 2] and the appendix of [24]; see also [24, Thm. 2.3].

In [65], (and also [66]), the authors show that certain cluster algebras can be associated to Riemann surfaces with boundary and a collection of marked points. Theorem 5.3.4 also holds in this context; see [65, Thm. 5.6]. See Chapter 8.

5.4 Root clusters

By Theorem 5.3.1, given any orientation Q of a Dynkin diagram (as a valued graph), there is a bijection $\alpha \mapsto x[\alpha]$ between the roots in the corresponding root system and the cluster variables of a cluster algebra of type Γ without coefficients. A *Q-root cluster C* is the set of roots corresponding to a cluster under this bijection. Thus we now consider how to obtain the Q-root clusters for a cluster algebra of finite type, mainly following [70].

Let Γ be a Dynkin diagram with corresponding root system Φ and simple system $\Delta = \{\alpha_1, \dots, \alpha_n\}$. The simple reflections s_{α_i} act on Φ, but this action does not restrict to $\Phi_{\geq -1}$. E.g. in type A_n, if $|i - j| = 1$ then $s_i(\alpha_i) = -\alpha_i - \alpha_j$. So, we adjust the action as follows. For $i \in I$, define $\sigma_i : \Phi_{\geq -1} \to \Phi_{\geq -1}$ by setting:

$$\sigma_i(\alpha) = \begin{cases} s_i(\alpha), & \text{if } s_i(\alpha) \in \Phi_{\geq -1}; \\ \alpha, & \text{else.} \end{cases}$$

Note that σ_i can also be defined as follows. We set $[\alpha : \alpha_i]$ to be the coefficient of α_i in the expansion of a root α in terms of the simple system $\alpha_1, \dots, \alpha_n$.

Then:

$$[\sigma_i(\alpha) : \alpha_j] = \begin{cases} [\alpha : \alpha_j], & \text{if } i \neq j; \\ -[\alpha : \alpha_i] - \sum_{k \neq i} a_{ik} \max([\alpha : \alpha_k], 0), & \text{if } i = j. \end{cases}$$

This definition makes sense for any $\alpha \in \mathbb{Z}\Phi$, the $\mathbb{Z}$-lattice spanned by Φ, but we note that the function σ_i is only piecewise linear, as opposed to the linearity of s_i. The σ_i are known as *truncated simple reflections*.

If Q is a valued quiver with underlying valued graph Γ, and k is a sink of Q, define $s_k(Q)$ to be the valued quiver obtained by reversing all of the arrows of Q incident with k.

Proposition 5.4.1. *[70, 124, 166] There is a unique collection of functions*

$$(-\|-)_Q : \Phi_{\geq -1} \times \Phi_{\geq -1} \to \mathbb{N} \cup \{0\},$$

for Q an orientation of Γ, such that

(a) $(\alpha\|\beta)_Q = (\sigma_k(\alpha)\|\sigma_k(\beta))_{s_k(Q)}$ whenever k is a sink or a source in Q, for all $\alpha, \beta \in \Phi_{\geq -1}$.

(b) $(-\alpha_i\|\beta)_Q = [\beta : \alpha_i]$ for all i and for all $\beta \in \Phi^+$;

(c) $(-\alpha_i\| - \alpha_j)_Q = 0$ for all i, j.

Proof. See [124, §4] and [166, §3], which build on [71, §3.1]. The argument in [124] is for the simply-laced case, and it is stated that a folding argument can be used for the non-simply-laced case; more details are given in [166, §3]. □

Remark 5.4.2. (a) The article [71] focuses on an alternating orientation of Γ (in which all vertices are sinks or sources), but this assumption is not necessary for the above result. In the alternating case, [71] shows that the compatibility degree can be interpreted in terms of a corresponding Y-system, a collection of Laurent polynomials associated to the elements of $\Phi_{\geq -1}$.

(b) In [166], the above is shown by interpreting σ_k in terms of a functor on the cluster category (i.e. in the representation-theoretic setting we have not focussed on here), induced by a BGP-reflection functor. Such functors were introduced in [16] as part of a proof of Gabriel's Theorem, classifying indecomposable modules over Dynkin quivers. They were also used in [124].

A subset of X of $\Phi_{\geq -1}$ satisfying $(\alpha \| \beta) = 0$ for all $\alpha, \beta \in X$ is said to be *Q-compatible*.

Theorem 5.4.3. *[167, §4] Let Q be an orientation of the valued graph Γ (of Dynkin type). Then the Q-root clusters are exactly the maximal Q-compatible subsets of $\Phi_{\geq -1}$.*

The alternating case was first considered in [71, §3]; see also [70, §3] (where the assumption of no coefficients is not needed). Also, one could use [28, Thm. 4.4] (which uses [23, Thm. 4.5]) for the simply-laced case.

5.5 Admissible sequences of sinks

We first recall the following.

Definition 5.5.1. (see [10, VII.5]) Let Q be a valued quiver. Then an *admissible sequence of sinks in Q* is a total ordering $v_1, \dots, v_n$ of the vertices of Q such that: v_1 is a sink in Q, v_2 is a sink in $s_{v_1}(Q), \dots, v_n$ is a sink in $s_{v_{n-1}} \cdots s_{v_1}(Q)$.

In general, a sequence $v_1, v_2, \dots, v_k$ of vertices of Q (of arbitrary length) satisfying the above conditions is said to be *adapted* to Q. The corresponding expression $s_{v_1} s_{v_2} \cdots s_{v_k}$ is also said to be adapted to Q. See [116, §4.7], where the case of adapted reduced expressions for w_0, the longest element of the Weyl group, is considered in the context of the canonical basis of a quantized enveloping algebra.

Thus, an admissible sequence of sinks is an adapted sequence of vertices of the quiver in which each vertex appears exactly once. We shall need the more general definition later, but for now focus on admissible sequences of sinks.

Lemma 5.5.2. *[10, VII.5.1] If $v_1 \dots v_n$ is an admissible sequence of sinks in a valued quiver Q then $s_{v_n} \cdots s_{v_1}(Q) = Q$.*

Proof. In the sequence $v_1, \dots, v_n$, each vertex appears exactly once. Each arrow, α, has exactly two vertices incident with it, say v_i and v_j. Then only s_{v_i} and s_{v_j} will affect α; each will reverse it, so that after applying $s_{v_n} \dots s_{v_1}$, Q will be returned to itself. □

Lemma 5.5.3. *[10, VII.5] Let Q be a finite acyclic valued quiver. Then Q has an admissible sequence of sinks.*

Proof. Suppose Q has no sink. Choose a vertex i_0. Since i_0 is not a sink, there is an arrow $i_0 \to i_1$, say. Repeating, we build a path $i_0 \to i_1 \to \dots$. Since Q is acyclic, there can be no repetitions in this sequence of vertices, a contradiction to the finiteness of Q. Hence Q has a sink.

We prove the result by induction on the number of vertices of Q. It is clear for a quiver with one vertex. Assume it to be true for a quiver with fewer than n vertices, where n is the number of vertices of Q. Let Q be a quiver with n vertices. By the above, Q has a sink, v_1. By induction, the full subquiver $Q \backslash \{v_1\}$ on Q with vertex v_1 removed has an admissible sequence $v_2, \dots, v_n$ of sinks. Then $v_1, v_2, \dots, v_n$ is an admissible sequence of sinks for Q: v_1 is a sink of Q by construction. In $s_{v_1}(Q)$, all arrows incident with v_1 have tail v_1. Then v_r is a sink in $s_{v_{r-1}}(Q) \dots s_{v_1}(Q) \backslash \{v_1\}$ by the induction hypothesis. If there is an arrow between v_1 and v_r, it must point towards v_r, since that is the case in $s_{v_1}(Q)$ and none of the $s_{v_2}, \dots, s_{v_{r-1}}$ change this arrow. Hence v_r is a sink in $s_{v_{r-1}}(Q) \dots s_{v_1}(Q)$ as required. □

Remark 5.5.4. The constructed sequence $v_1, \dots, v_n$ has the following properties:

(a) v_r is a sink in $Q \backslash \{v_1, \dots, v_{r-1}\}$ for $r = 1, \dots, n$.

(b) If $v_i \to v_j$ is an arrow in Q then $i > j$.

The second property follows easily using induction on r. We note that in fact (a) is equivalent to $v_1, \dots, v_r$ being an admissible sequence of sinks. Similarly for (b).

Now, suppose that Q is an orientation of a Dynkin diagram Γ (as valued quiver). Then Q is acyclic, so by Lemma 5.5.3, it has an admissible sequence of sinks, $v_1, v_2, \dots, v_n$, satisfying $v_i \to v_j$ implies $i > j$.

For example,

$$v_3 \to v_2 \to v_1, \qquad v_3 \to v_1 \leftarrow v_2.$$

By Lemma 5.5.2, we have $s_{v_n} \dots s_{v_1}(Q) = Q$. The element $c_Q = s_{v_n} \dots s_{v_1} \in W_\Gamma$, the corresponding reflection group, is a Coxeter element, and is said to be *adapted* to Q. (Note that conversely, if $c = s_{v_n} \cdots s_{v_1}$ is any Coxeter element, it is adapted to the orientation on the Dynkin diagram with the property that, whenever v_i, v_j are linked by an edge, the edge is oriented towards the lower of i, j.) Let $\sigma_Q = \sigma_{v_n} \dots \sigma_{v_1}$ be the corresponding piecewise-linear map.

Lemma 5.5.5. *[70, 166] Each σ_Q-orbit on $\Phi_{\geq -1}$ intersects $\{-\alpha_1, \dots, -\alpha_n\}$.*

Proof. In the generality we are working in, this follows from [166, Thm. 3.4]. A similar result was first proved in the alternating case in [71, Thm. 2.6]. □

This gives a method for computing $(-\|-)_Q$ without referring to $(-\|-)_{Q'}$ for other orientations Q'. By Proposition 5.4.1 we have:

$$\begin{aligned}(\alpha\|\beta)_Q &= (\sigma_Q(\alpha)\|\sigma_Q(\beta))_{s_{v_n}\dots s_{v_1}(Q)}\\ &= (\sigma_Q(\alpha)\|\sigma_Q(\beta))_Q.\end{aligned}$$

We repeatedly apply this until we get $\sigma_Q^r(\alpha) = -\alpha_i$ for some i on the left-hand-side. If $\sigma_Q^r(\beta) \in \Phi^+$, we obtain

$$(\alpha\|\beta)_Q = [-\alpha_i : \sigma_Q^r(\beta)],$$

again using Proposition 5.4.1. Otherwise, we get zero.

This is similar to [71, §3], for the alternating case, as we now explain. Since Γ is a tree, we can divide its vertices, I, into two subsets, I_+ and I_-, such that all edges link a vertex in I_+ with a vertex in I_-. Since Q is alternating we have that (up to a choice of sign), each vertex in I_+ is a sink. Let $c_+ = \prod_{i\in I_+} s_i$ and $c_- = \prod_{i\in I_-} s_i$. Then

$$c_Q = c_-c_+.$$

So we have

$$(\alpha\|\beta)_Q = (\sigma_-\sigma_+(\alpha)\|\sigma_-\sigma_+(\beta))_Q,$$

for all α, β. In fact, by [71, §3] we have

$$(\alpha\|\beta)_Q = (\sigma_+(\alpha)\|\sigma_+(\beta))_Q$$

and

$$(\alpha\|\beta)_Q = (\sigma_-(\alpha)\|\sigma_-(\beta))_Q$$

for all $\alpha, \beta \in \Phi_{\geq -1}$.

Example 5.5.6. (Type A_3) Let Q be the quiver

$$1 \longrightarrow 2 \longleftarrow 3,$$

so $c_Q = s_1s_3s_2$. Then we have:

$$\begin{aligned}(\alpha_1+\alpha_2\|\alpha_2)_Q &= (s_1s_3s_2(\alpha_1+\alpha_2)\|s_1s_3s_2(\alpha_2))_Q\\ &= (-\alpha_1\|-\alpha_2)_Q\\ &= 0,\end{aligned}$$

as expected, since $\alpha_1+\alpha_2, \alpha_2$ are in the same Q-root cluster. We also have:

$$\begin{aligned}(\alpha_1+\alpha_2+\alpha_3\|\alpha_2+\alpha_3)_Q &= (s_1s_3s_2(\alpha_1+\alpha_2+\alpha_3)\|s_1s_3s_2(\alpha_2+\alpha_3))_Q\\ &= (\alpha_2\|-\alpha_3)_Q\\ &= (s_1s_3s_2(\alpha_2)\|s_1s_3s_2(-\alpha_3))_Q\\ &= (-\alpha_2\|\alpha_3)_Q\\ &= 0.\end{aligned}$$

Note that $(-\alpha_1 \| \alpha_1 + \alpha_2)_Q = 1$, so non-zero values can occur. Also, if Γ is simply-laced we have

$$(\alpha \| \beta)_Q = (\beta \| \alpha)_Q, \text{ for all } \alpha, \beta \in \Phi_{\geq -1}.$$

5.6 Number of clusters

A Q-root cluster is said to be *positive* if it consists entirely of positive roots. The corresponding cluster itself is also said to be positive. Note that this depends on a choice of Q.

Theorem 5.6.1. *Let Γ be a Dynkin diagram and $\mathcal{A}$ a cluster algebra of type Γ. Recall that we have denoted the exponents of Γ by e_i, and its Coxeter number by h. We have:*

(a) [71, Thm. 1.9] and [70, §2] The number of seeds (or root clusters) in $\mathcal{A}$ is given by

$$\prod_{i=1}^{n} \frac{e_1 + h + 1}{e_i + 1}.$$

(b) [71, Prop. 3.9] and [70, §2]; [124, §6] The number of positive root clusters (for some orientation Q of Γ) in $\mathcal{A}$ is given by

$$\prod_{i=1}^{n} \frac{e_1 + h - 1}{e_i + 1}.$$

Note that in (b), the result in [71] is for Q alternating, and that the proof in [124] uses Proposition 5.4.1.

In the table below, we give the number of clusters (denoted N_C) and the number (denoted N_C^+) of positive clusters in each type. These numbers appear in [71].

Type	A_n	B_n/C_n	D_n	E_6	E_7	E_8	F_4	G_2
N_C	$\frac{1}{n+2}\binom{2n+2}{n+1}$	$\binom{2n}{n}$	$\frac{3n-2}{n}\binom{2n-2}{n-1}$	833	4160	25080	105	8
N_C^+	$\frac{1}{n+1}\binom{2n}{n}$	$\binom{2n-1}{n}$	$\frac{3n-4}{n}\binom{2n-3}{n-1}$	418	2431	17342	66	5

The sequence of numbers with nth term $\frac{1}{n+2}\binom{2n}{n}$ is known as the sequence of *Catalan numbers*, and arises in many combinatorial contexts. For more information about this, see [157].

6 Generalized Associahedra

Associated to each cluster algebra of finite type (or Dynkin diagram) is a corresponding abstract simplicial complex on the clusters and, by a result of [35], a corresponding polytope, known as the *generalized associahedron*. We give an overview of these results in this chapter.

6.1 Fans

We first give a short introduction to the notion of a fan, following [171].

A *cone* is a subset κ of a Euclidean space V, non empty and closed under non-negative linear combinations. An example of a cone is a subset κ of the form

$$\kappa = \{\lambda_1 v_1 + \cdots + \lambda_k v_k : \lambda_1 \geq 0, \ldots, \lambda_k \geq 0\},$$

which is called a *polyhedral cone*. A subset κ of V is a polyhedral cone if and only if it is an intersection of closed half-spaces. A *face* of a polyhedral cone κ is a subset of κ of the form

$$\text{face}_{\omega}(\kappa) = \{x \in \kappa : \omega(x) = 0\},$$

where ω is a linear functional on V such that $\omega(y) \leq 0$ for all $y \in \kappa$. Figure 6.1 illustrates a polyhedral cone in $\mathbb{R}^2$. The two bounding half-lines and the origin are all faces. Note that a cone is always a face of itself, taking $\omega = 0$. All other faces are known as *proper faces*.

As an example of a non-polyhedral cone, we could take the cone generated by $\{(a, a^2) : a \in \mathbb{Z}_{\geq 0}\}$ in $\mathbb{R}^2$.

A *fan* in V is a family $\mathcal{F} = \{\kappa_1, \ldots, \kappa_N\}$ of polyhedral cones such that:

(a) Every non-empty face of a cone in $\mathcal{F}$ is a cone in $\mathcal{F}$,

(b) The intersection of any two cones in $\mathcal{F}$ is a face of both.

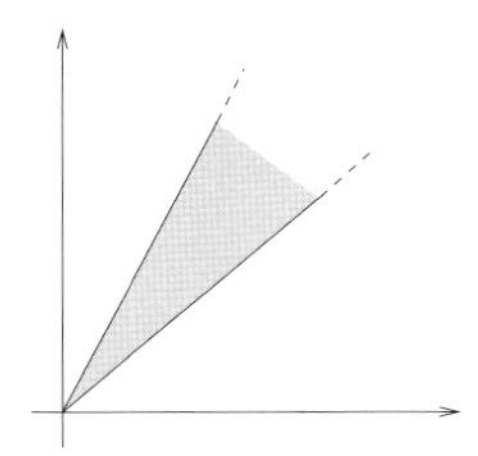

Figure 6.1. A polyhedral cone in $\mathbb{R}^2$.

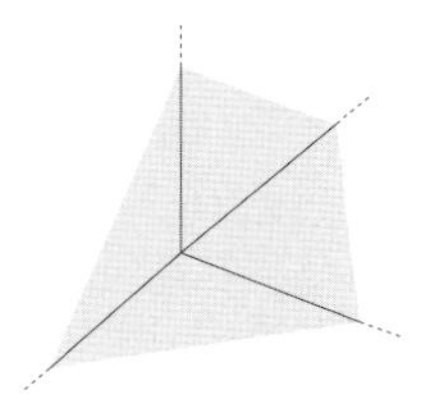

Figure 6.2. Example of a fan in $\mathbb{R}^2$.

Figure 6.2 gives an example of a fan in $\mathbb{R}^2$ containing 4 cones of maximal dimension (shaded).

A fan is said to be *complete* if $\bigcup_{i=1}^{N} \kappa_i = V$. It is said to be *simplicial* if all of its cones are simplicial, i.e. spanned by linearly independent vectors.

6.2 The cluster fan

Let $J \subseteq I$. By Proposition 4.3.2,

$$\Phi|_{I \setminus J} = \{\sum_{i=1}^{n} d_i \alpha_i \ : \ d_i = 0 \text{ for all } i \notin J\}$$

is again a root system, with simple system given by the simple roots α_i for $i \notin J$.

We define the *negative support* of a Q-root cluster C to be the set

$$\{i \in I \ : \ -\alpha_i \in C\}.$$

Lemma 6.2.1. *[71, Prop. 3.6] There is a bijection between the following two sets:*

$$\begin{array}{ccc} \left\{ \begin{array}{c} Q\text{-root clusters with} \\ \text{negative support containing } J \end{array} \right\} & \longleftrightarrow & \left\{ \begin{array}{c} Q|_{I \setminus J}\text{-root} \\ \text{clusters} \end{array} \right\} \\ C & \longmapsto & C \setminus \{-\alpha_i \ : \ i \in J\}, \end{array}$$

where $Q|_{I \setminus J}$ indicates the full subquiver of Q on vertices $I \setminus J$.

Proof. This is essentially [71, Prop. 3.6]; see also [124, Prop. 3.4]. Let C be a Q-root cluster with negative support containing J. Note that, by the definition of compatibility degree, no positive root $\alpha \in C$ can have a root α_i, $i \in J$, in its expansion. So the set $C \setminus \{-\alpha_i \mid i \in J\}$ lies in $(\Phi|_{I \setminus J})_{\geq -1}$.

Note that s_i', the simple reflection in the hyperplane orthogonal to α_i in $\text{span}(\Phi|_{I \setminus J})$, and the simple reflection s_i coincide on $\Phi|_{I \setminus J}$, so the same is true for σ_i and σ_i'. It then follows from the definition of compatibility degree (c.f. Proposition 5.4.1) that, for all $\alpha, \beta \in (\Phi|_{I \setminus J})_{\geq -1}$,

$$(\alpha||\beta)_{Q|_{I\setminus J}} = (\alpha||\beta)_Q. \tag{6.1}$$

Thus $C \setminus \{-\alpha_i : i \in J\}$ is a $Q|_{I\setminus J}$-root cluster. Furthermore, any such root cluster can be extended to a Q-root cluster by forming the union with $\{-\alpha_i : i \in J\}$. It is clear that these two maps are inverses to each other, so we have a bijection as claimed. □

Corollary 6.2.2. *[71, Prop. 3.6] The map in Lemma 6.2.1 induces bijections between the following sets:*

$$\left\{\begin{matrix} Q\text{-compatible subsets of } \Phi_{\geq -1} \\ \text{with support containing } J \end{matrix}\right\} \longleftrightarrow \left\{\begin{matrix} Q|_{I\setminus J}\text{-compatible subsets} \\ \text{of } (\Phi_{I\setminus J})_{\geq -1} \end{matrix}\right\}$$

and

$$\left\{\begin{matrix} Q\text{-compatible subsets of } \Phi_{\geq -1} \\ \text{with support } J \end{matrix}\right\} \longleftrightarrow \left\{\begin{matrix} \text{positive } Q|_{I\setminus J}\text{-compatible subsets} \\ \text{of } (\Phi_{I\setminus J})_{\geq -1} \end{matrix}\right\}$$

Theorem 6.2.3. *Let Q be an orientation of the Dynkin diagram Γ (as valued graph). Then each root cluster of type Q forms a $\mathbb{Z}$-basis of $\mathbb{Z}\Phi$ (the root lattice), and the cones spanned by compatible subsets of roots form a complete simplicial fan in $\mathbb{R}\Phi = V$, known as the* cluster fan.

Proof. We adapt the proof from [71, Thm. 1.10], which considers the case where Q is alternating. We start with the $\mathbb{Z}$-basis claim. Recall that I is the set of vertices of Γ (hence also of Q).

To show that every Q-root cluster is a $\mathbb{Z}$-basis of $\mathbb{Z}\Phi$, we proceed by induction on the rank of Φ (note that we don't assume that Φ is irreducible). The result clearly holds for rank 1. For the general case, we assume it holds for smaller rank. If C contains any negative roots then, by Lemma 6.2.1, $C \setminus \{-\alpha_i : i \in J\}$, where J is the negative support of C, is a $\mathbb{Z}$-basis for $\mathbb{Z}\Phi_{I\setminus J}$, so C is a $\mathbb{Z}$-basis for $\mathbb{Z}\Phi$.

If all of the roots in C are positive and $s_i(C)$ has the same property, then σ_i and s_i coincide on C, and since s_i is an orthogonal transformation, it preserves the property of being a $\mathbb{Z}$-basis. Let $j_1, j_2, \dots, j_k$ be a sequence of elements of I adapted to Q (see Definition 5.5.1) such that $s_{j_k} s_{j_{k-1}} \cdots s_{j_1} Q = Q_{alt}$, an alternating orientation of Γ. Using this sequence, together with Lemma 5.5.5, we see that there is a sequence $j_1, j_2, \dots, j_l$ of elements of I adapted to Q such that $\sigma_{j_l} \cdots \sigma_{j_1}(C)$ contains a negative simple root. We choose such a sequence with a minimal such l.

Since the sets

$$C, \sigma_{j_1}(C), \dots, \sigma_{j_{l-1}} \cdots \sigma_{j_1}(C)$$

consist only of positive roots, σ_{j_r} and s_{j_r} coincide on $\sigma_{j_{r-1}} \cdots \sigma_{j_1}(C)$ for $r = 1, 2, \dots, l$. Hence we have

$$\sigma_{j_l} \cdots \sigma_{j_1}(C) = s_{j_l} \cdots s_{j_1}(C).$$

Since $s_{j_k} \cdots s_{j_1}(C)$ is a Q-root cluster and contains a negative simple root, it follows from the above that it is a $\mathbb{Z}$-basis for $\mathbb{Z}\Phi$. It follows that C is also such a basis (and thus also that C is an $\mathbb{R}$-basis for $\mathbb{R}\Phi$). Hence, in particular, the nonnegative real span of any Q-compatible subset of roots is a simplicial cone.

For a element $\gamma \in \mathbb{R}\Phi$ written as $\gamma = \sum_{i \in I} d_i \alpha_i$, let $S(\gamma) = \{i \in I : d_i \neq 0\}$, $S_+(\gamma) = \{i \in I : d_i > 0\}$ and $S_-(\gamma) = \{i \in I : d_i < 0\}$. To prove the completeness of the fan, we show that each $\gamma \in \mathbb{R}\Phi$ has a unique *cluster expansion*, i.e. an expression of the form

$$\gamma = \sum_{\alpha \in \Phi_{\geq -1}} m_\alpha \alpha,$$

where all roots α appearing are Q-compatible and the m_α are real numbers.

We make the following claim:

Claim (a): If $\alpha \in \Phi_{\geq -1}$ appears in a cluster expansion of γ, then either $\alpha \in \Phi^+$ and $S(\alpha) \subseteq S_+(\gamma)$ or $\alpha = -\alpha_i$ for some $i \in S_-(\gamma)$.

Proof of claim: If $\alpha \in \Phi^+$ appears in a cluster expansion of γ, then no $-\alpha_i$ for $i \in S(\alpha)$ can appear in the expansion (as it is not compatible with α), so $S(\alpha) \subseteq S_+(\gamma)$. If $\alpha = -\alpha_i$ for some i appears in a cluster expansion of γ then similarly no $\alpha \in \Phi^+$ with $i \in S(\alpha)$ can appear in the expansion, and the claim is proved.

Write $\gamma = \sum_{i \in I} d_i \alpha_i$ and let $\gamma^+ = \sum_{i \in S_+(\gamma)} d_i \alpha_i$ be the positive part of γ.

Claim (b): An element $\gamma \in \mathbb{R}\Phi$ has a unique cluster expansion if and only if γ^+ has a unique cluster expansion in $\Phi|_{S_+(\gamma)}$. Since the negative simple roots do not affect the existence of a cluster expansion, this follows from Claim (a) and equation (6.1) above.

This means it is enough to restrict to the case $\gamma \in \mathbb{R}_{\geq 0}\Phi$. Note then that only positive roots in the cluster appear in the cluster expansion of γ. By Claim (a), the result holds for $\gamma = 0$, so assume $\gamma \neq 0$. Again, we use induction on the rank.

Since $w_0(\alpha_i)$ is negative for each i (see Proposition 4.10.1 and the comment afterwards), $w(\gamma) \notin \mathbb{R}_{\geq 0}\Phi$. By [116, Prop. 4.12], there is a reduced expression for w_0 adapted to Q (see the paragraph after Definition 5.5.1). It follows that there is a sequence $j_1, j_2, \ldots, j_l$ of elements of I adapted to Q such that $\sigma_{j_l} \cdots \sigma_{j_1}(\gamma)$ is not a nonnegative combination of simple roots. Let l be minimal with this property. Since σ_i and s_i coincide on positive roots, we see that $\gamma = \sum_{\alpha \in \Phi^+} m_\alpha \alpha$ is a Q-cluster expansion of γ if and only if $s_i(\gamma) = \sum_{\alpha \in \Phi^+} m_\alpha s_i(\alpha)$ is an $s_i(Q)$-cluster expansion of $s_i(\gamma)$, using Proposition 5.4.1. So γ has a unique Q-cluster expansion if and only if $s_i(\gamma)$ has a unique $s_i(Q)$-cluster expansion.

Repeating this argument and noting that $\sigma_{j_r} \cdots \sigma_{j_1}(\gamma)$ is a nonnegative combination of simple roots for $r = 1, 2, \ldots, l-1$, we see that γ has a unique Q-cluster expansion if and only if $s_{j_l} \ldots s_{j_1}(\gamma)$ has a unique $s_{j_l} \ldots s_{j_1}(Q)$-cluster expansion. So γ has a unique Q-cluster expansion if and only if $s_{j_l} \ldots s_{j_1}(\gamma)$ has a unique $s_{j_l} \ldots s_{j_1}(Q)$-cluster expansion.

By Claim (b), this holds if and only if $(s_{j_l} \ldots s_{j_1}(\gamma))^+$ has a unique $s_{j_l} \ldots s_{j_1}(Q)$-cluster expansion. But $(s_{j_l} \ldots s_{j_1}(\gamma))^+$ lies in a root lattice of smaller rank, so this holds by the inductive hypothesis and we are done.

Consider two cones $\mathbb{R}_{\geq 0}C$ and $\mathbb{R}_{\geq 0}C'$, where C and C' are compatible subsets of roots. Then we have

$$\mathbb{R}_{\geq 0}(C \cap C') \subseteq \mathbb{R}_{\geq 0}C \cap \mathbb{R}_{\geq 0}C'.$$

If $v \in \mathbb{R}_{\geq 0}C \cap \mathbb{R}_{\geq 0}C'$, then by the uniqueness of its cluster expansion, above, it lies in $\mathbb{R}_{\geq 0}(C \cap C')$ so we have equality and we see that the intersection of any two cones in the collection is another such cone, i.e. that the cones spanned by compatible subsets of roots form a fan in V.

The existence of cluster expansions implies that the union of the cones $\mathbb{R}_{\geq 0}C$ contains $\mathbb{Z}\Phi$. Since this union is closed under addition and under multiplication by positive real numbers, it must be the whole of $\mathbb{R}\Phi$, so the fan is complete and we are done. □

Note that if $\gamma \in \mathbb{Z}\Phi$, it has a unique cluster expansion by Theorem 6.2.3. Since the roots appearing in the expansion are part of a $\mathbb{Z}$-basis for $\mathbb{Z}\Phi$, again by Theorem 6.2.3, the coefficients in the cluster expansion of γ must all be integers (Alternatively, the above argument could be used directly to show this, by working in the root lattice rather than $\mathbb{R}\Phi$).

6.3 The cluster complex

The *convex hull* of a set of points $x_1, \ldots, x_r$ in a Euclidean space V is the smallest convex set containing $x_1, \ldots, x_r$, i.e.

$$\left\{ \sum_{i=1}^{r} \lambda_i x_i \;\middle|\; \sum_{i=1}^{r} \lambda_i = 1, \lambda_i \geq 0 \text{ for all } i \right\}.$$

An *n-simplex* is the convex hull of a set S of $n+1$ points in some $\mathbb{R}^n$ which lie in general position, i.e. no subset of S of cardinality k lies in an affine subspace of dimension $k-2$. For example, no three points are collinear, no four points lie in a plane, etc.

Then a 0-simplex is a point and a 1-simplex is a single edge between two points. A 2-simplex is a triangle and a 3-simplex is a tetrahedron.

A *face* of an n-simplex is the convex hull of a subset of its points.

A *Euclidean simplicial complex* is a set $\mathcal{S}$ of simplices in a Euclidean space V such that

(a) If $S \in \mathcal{S}$ then every face of S lies in $\mathcal{S}$;

(b) The intersection of any two simplices in $\mathcal{S}$ is empty or a face of both simplices;

(c) Every point in a simplex in $\mathcal{S}$ has a neighborhood which only finitely many simplices intersect.

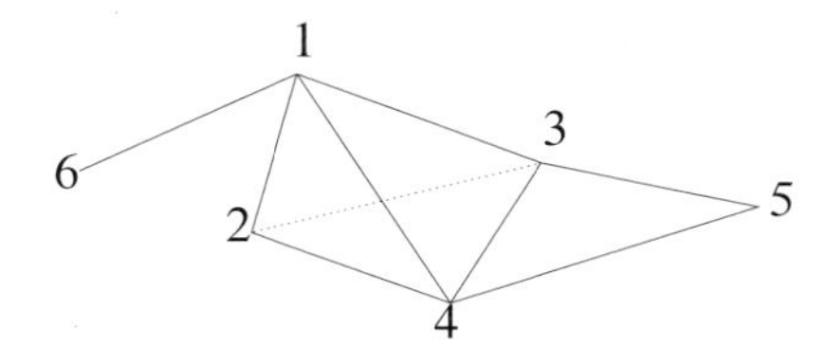

Figure 6.3. An example of a Euclidean simplicial complex.

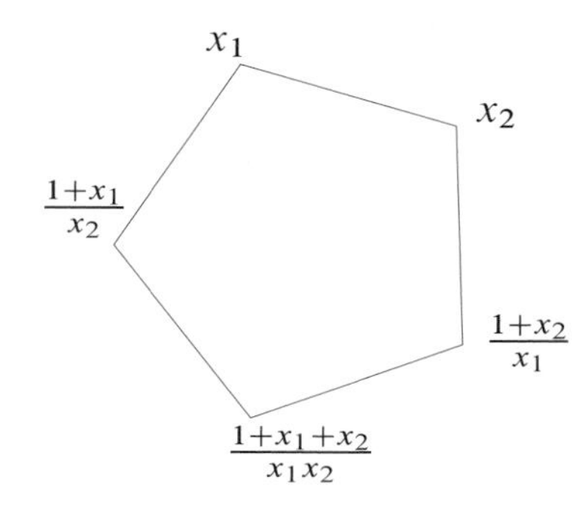

Figure 6.4. A geometric realization of the type A_2 cluster complex.

A Euclidean simplicial complex inherits the real topology from V. For an example in three dimensions, see Figure 6.3.

An *abstract simplicial complex* is a finite collection $\mathcal{K}$ of sets such that if $K \in \mathcal{K}$ and $K' \subseteq K$, $K' \neq \emptyset$, then $K' \in \mathcal{K}$.

Such an abstract simplicial complex has a corresponding *realization* as a Euclidean simplicial complex. Let $B = \cup_{K \in \mathcal{K}} K$ and take $V = \mathbb{R}^N$ where $N = |B|$, with standard basis $e_b, b \in B$. Then each $K \in \mathcal{K}$ corresponds to a simplex in V given by the convex hull of $\{e_b : b \in K\}$. For example, the geometric realization of the abstract simplicial complex:

$$\mathcal{P}([1,4]) \cup \mathcal{P}([1,6]) \cup \mathcal{P}([3,5]) \backslash \{\emptyset\},$$

where $\mathcal{P}$ denotes the power set of a set, is given in Figure 6.3.

The *cluster complex* of type Γ (where Γ is a Dynkin diagram) is the abstract simplicial complex whose simplices are the subsets of clusters, or, equivalently, the Q-compatible sets of almost positive roots for some orientation Q of Γ. For example, in type A_2, a geometric realisation of the cluster complex is a pentagon; see Figure 6.4.

We have the following corollary of Theorem 6.2.3.

Corollary 6.3.1. *[71, Cor. 1.11] The cluster complex associated to any Dynkin diagram is homeomorphic to a sphere of dimension $n-1$.*

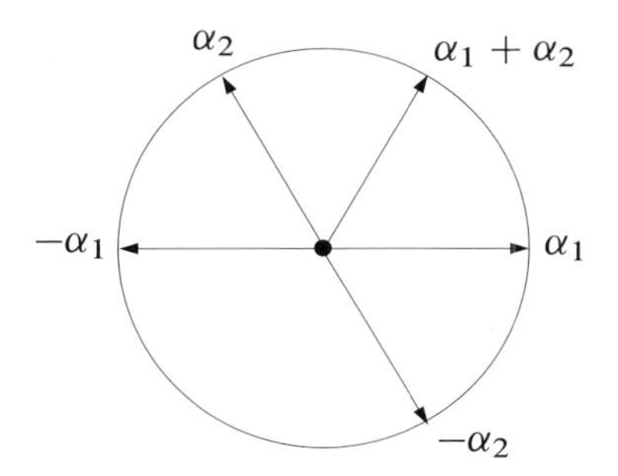

Figure 6.5. The circle here is a geometric realization of the type A_2 cluster complex in $\mathbb{R}\Phi$. The cones spanned by the position vectors of adjacent roots on the circle are the maximal cones in the cluster fan.

Proof. Each maximal simplex, which is a maximal Q-compatible set of almost positive roots, corresponds to the intersection of a sphere centered at the origin of $\mathbb{R}\Phi$ with a maximal cone in the cluster fan. This gives rise to a geometric realization of the cluster complex which is homeomorphic to a sphere of dimension $n - 1$. □

The type A_2 case is illustrated in Figure 6.5: in this case the cluster complex is homeomorphic to a sphere of dimension 1, i.e. a circle.

We remark that the article [63] defines a generalized cluster complex for each integer $m \geq 1$. For more information on cluster complexes from a representation-theoretic perspective, see [96] and references therein.

6.4 Normal fans

For more information on the topics discussed in this section, see [159] and [171]. Let P be a (convex) polytope in V, i.e. the convex hull of a finite set of points. A linear functional ω on V defines a subset

$$\mathrm{face}_\omega(P) = \{x \in P : \omega(x) \geq \omega(p), \text{ for all } p \in P\}.$$

Such subsets are defined to be the *faces* of P. Note that the same definition, used for a polyhedral cone, coincides with the definition of face given in Section 6.1. As in that case, a polytope is always a face of itself; all other faces are known as *proper* faces.

Linear functionals on V are of the form $(y, -)$, where $(-, -)$ is the bilinear form on V and y is an element of V. *Level sets* of the linear functional $(y, -)$ are the affine hyperplanes which are translations of the hyperplane orthogonal to y in V. So a proper face can be thought of as where a moving hyperplane crossing P leaves P.

For a face F, the *normal cone* of F is the cone:

$$\kappa_F = \{y \in V : \mathrm{face}_{(y,-)}(P) = F\}.$$

If F, F' are faces of P, then F' is a face of F if and only if κ_F is a face of $\kappa_{F'}$. The normal cones of the faces of non-empty polytope P form a fan, with maximal cones corresponding to the vertices of P. This is referred to as the *normal fan* of P.

For an illustration in 2 dimensions, see Figure 6.6, from [35, §1]. This shows a convex polytope P whose normal fan is the Q-cluster fan where Q is an orientation of the Dynkin diagram of type A_2. For a vertex v, a vector y lies in κ_v if and only if $\mathrm{face}_{(y,-)}(P) = \{v\}$, i.e. if and only if

$$\{x \in P : (y, x) \geq (y, p), \text{ for all } p \in P\} = \{v\}.$$

This holds if and only if the affine hyperplane (line) $\{x \in V : (y, x) = (y, v)\}$ intersects P only in v, and y, when translated to v, lies on the side of this line not containing P. This happens if and only if the vector y, when translated to v, lies between the perpendiculars to the sides of P incident with v. Thus the maximal cones have been drawn based at the corresponding vertices of P. Translating the cones back to the origin we obtain the normal fan of P, which coincides with the cluster fan of type A_2 illustrated in Figure 6.5.

6.5 The generalized associahedron

Recall that a polytope is *simple* if every vertex lies in precisely n facets (i.e. faces of maximal dimension).

Theorem 6.5.1. *[35, Thm. 1.4] Let Q_{alt} be an alternating orientation of a Dynkin diagram Γ (as valued quiver). Then the corresponding Q_{alt}-cluster fan is the normal fan of a simple n-dimensional convex polytope P_Γ.*

We have the following correspondences:

$$\begin{gathered}\{Q_{alt}\text{-compatible sets of roots}\}\\ \updownarrow\\ \{\text{simplices of the cluster complex}\}\\ \updownarrow\\ \{\text{cones in the } Q_{alt}\text{-cluster fan}\}\\ \updownarrow\\ \{\text{faces of } P_\Gamma\}\end{gathered}$$

The first two correspondences preserve inclusion, while the last reverses inclusion. The correspondences restricts to correspondences between Q_{alt}-root clusters, maximal simplices of the cluster complex, maximal cones in the Q_{alt}-cluster fan and vertices of P_Γ.

The 1-skeleton of the polytope P_Γ is isomorphic to the exchange graph. The face poset of the polytope is isomorphic to the poset of simplices in the cluster complex under reverse inclusion. The polytope $\mathcal{P}_\Gamma$ is known as a *generalized associahedron*. For more information on generalized associahedra, see the lecture notes [64] or the survey [92].

7 Periodicity

In this chapter we consider two aspects of periodicity in cluster algebras. First we consider quivers which have periodicity properties with respect to mutation, and the integer recurrences corresponding to them, including a classification result from [75]. Secondly we consider the categorical periodicity considered by B. Keller in his proof [103] of a conjecture of Al. B. Zamolodchikov in the context of the thermodynamic Bethe Ansatz.

7.1 The Somos-4 recurrence

Recall Example 2.3.3(ii). We start with the quiver S_4:

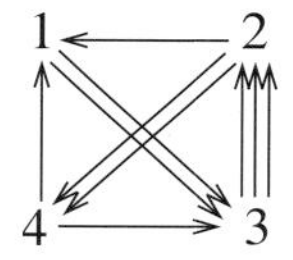

and mutating at vertex 1, we obtain $\mu_1 S_4$:

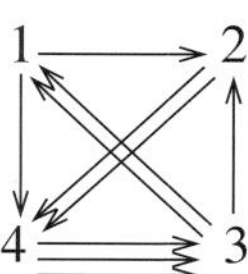

which is just S_4 rotated through $\pi/2$ clockwise. The corresponding exchange relation is:

$$x_1' x_1 = x_2 x_4 + x_3^2.$$

Let us denote x_1' by x_5, so that we have:

$$x_5 x_1 = x_2 x_4 + x_3^2,$$

and

$$\begin{aligned}\mu_1\big(\{x_1, x_2, x_3, x_4\}, S_4\big) &= \big(\{x_5, x_2, x_3, x_4\}, \mu_1 S_4\big)\\ &= \big(\{x_2, x_3, x_4, x_5\}, S_4\big).\end{aligned}$$

where in the last seed, the variables x_2, x_3, x_4, x_5 correspond to the vertices $1, 2, 3, 4$ in the diagram of S_4 displayed above. Mutating at x_2, we get

$$(\{x_3, x_4, x_5, x_6\}, S_4),$$

where $x_2x_6 = x_3x_5 + x_4^2$. Continuing the mutation in this way, we see that we obtain the terms of the recurrence

$$x_kx_{k+4} = x_{k+1}x_{k+3} + x_{k+2}^2, \qquad k = 1, 2, \ldots,$$

i.e. the recurrence defining the Somos-4 sequence (see Section 1.3). Specializing x_1, x_2, x_3, x_4 to 1, we obtain the Somos-4 sequence. The following theorem is key.

Theorem 7.1.1. *[The Laurent Phenomenon] [67, Thm. 3.1]. Let $\mathcal{A}$ be a cluster algebra defined by initial cluster $(\widetilde{\mathbf{u}}, \widetilde{B})$ where $\widetilde{\mathbf{u}} = \{u_1, \ldots, u_n, \ldots, u_m\}$, $n \leq m$ (thus $u_{n+1}, \ldots, u_m$ are the coefficients). Then, any cluster variable u of $\mathcal{A}$ can be written as a Laurent polynomial in $u_1, \ldots, u_n$ with coefficients which are integer polynomials in $u_{n+1}, \ldots, u_m$.*

Note that, a priori, u is only a rational function in $u_1 \ldots, u_m$.

Corollary 7.1.2. *[68, Ex. 3.3] The kth term in the sequence (x_k) defined above is a Laurent polynomial in $x_1, \ldots, x_4$ with integer coefficients. Thus, the kth term in the Somos-4 sequence is an integer.*

A sequence with the property in Corollary 7.1.2 is said to be a *Laurent sequence.*

In the proof in [68] (which builds on unpublished work of N. Elkies and J. Propp), a higher dimensional recurrence known as the *octahedron recurrence* is first shown to be Laurent, and then a folding procedure is used. In [155], formulas for the terms are given in terms of perfect matchings, giving a positivity result. A bounded version of the octahedron recurrence, introduced in [91], is studied in [90], again with results obtained in terms of perfect matchings, and shown to be periodic. See also Section 1.3 for more information.

The above approach is from [75, Ex. 8.4], which gives a reinterpretation of the proof in [68].

7.2 Period 1 quivers

The above procedure will work for any quiver satisfying

$$\mu_1(Q) = \rho(Q),$$

where ρ is the permutation $(1\ 2 \cdots n)$. Here, in $\rho(Q)$, the number of arrows from i to j is the number of arrows from $\rho^{-1}(i)$ to $\rho^{-1}(j)$ in Q. Call a quiver a *period* 1 quiver if it satisfies this equation. It is often helpful to regard such quivers as being embedded in a disk, with the vertices $1, 2, \ldots, n$ equally spaced clockwise around the boundary.

Fix $1 \leq t \leq \frac{n}{2}$. Let $P_n^{(t)}$ be the quiver with a single edge joining each vertex $i \in I$ to vertex $i + t \mod n$ (taking a representative in $1, \ldots, n$) for each i. The edge is

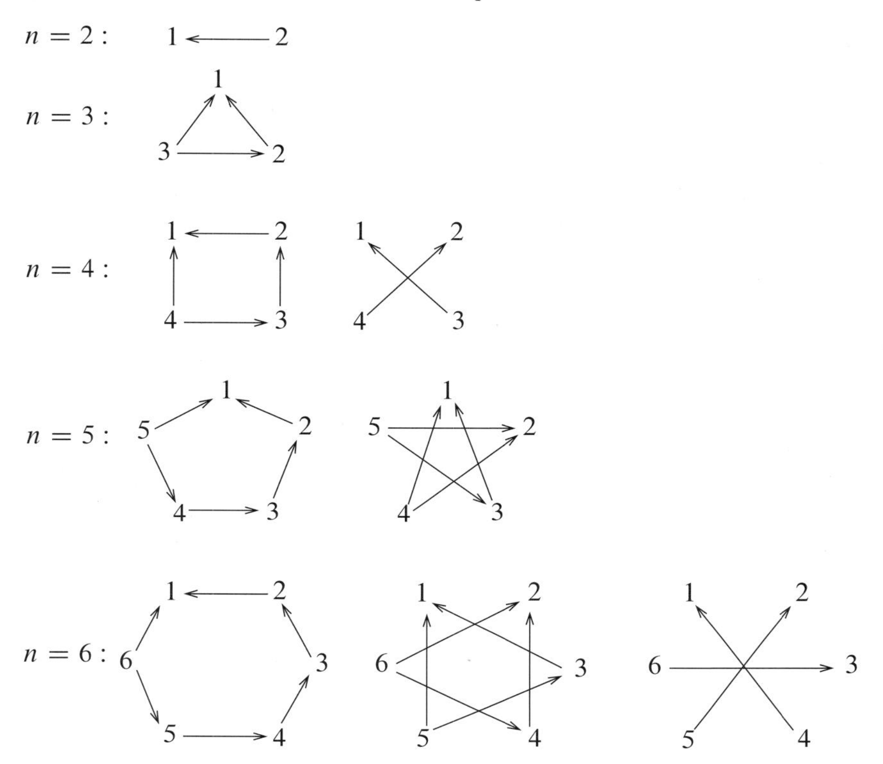

Figure 7.1. Primitive quivers of degree $1, 2 \ldots 6$.

oriented towards the end-point with the smaller label. The quiver $P_n^{(t)}$ is called the tth *primitive* quiver of degree n.

The primitive quivers for $n = 1, 2, \ldots, 6$ are shown in Figure 7.1.

Given two quivers Q, Q' with vertices $1, \ldots, n$ let $Q + Q'$ denote the quiver for which the number of arrows from i to j equals the number of arrows from i to j in Q plus the number of arrows from i to j in Q', for any pair i, j of vertices. Similarly, define rQ for any integer r.

Theorem 7.2.1. *[75, Thm. 6.6] Any period* 1 *quiver has the following form. Let* $r = \lfloor \frac{n}{2} \rfloor$ *and let* $u_1, \ldots, u_r \in \mathbb{Z}$. *For all* $i, j \in [1, r]$, *set*

$$\begin{aligned} \varepsilon_{ij} &= \frac{1}{2}(u_i|u_j| - u_j|u_i|) \\ &= \begin{cases} \operatorname{sign}(u_j)u_iu_j & \text{if } u_i, u_j \text{ have opposite sign;} \\ 0 & \text{else.} \end{cases} \end{aligned}$$

Then set

$$Q = \sum_{t=1}^{r} u_t P_n^{(t)} + \sum_{l=1}^{r-1} \sum_{t=1}^{r-l} \varepsilon_{l,l+t} \rho^l (P_{n-2l}^{(t)}).$$

The second term is a sum of primitives with fewer vertices, rotated appropriately by ρ and can be regarded as a correction term.

For example, $S_4 = P_4^{(1)} - 2P_4^{(2)} + 2\rho P_2^{(1)}$ which is equal to:

$$\begin{array}{ccc} 1 & \longleftarrow & 2 \\ \uparrow & & \uparrow \\ 4 & \longrightarrow & 3 \end{array} - 2 \left(\begin{array}{ccc} 1 & & 2 \\ & \nwarrow \nearrow & \\ & \times & \\ 4 & & 3 \end{array} \right) + 2 \left(\begin{array}{ccc} 1 & & 2 \\ & & \uparrow \\ 4 & & 3 \end{array} \right).$$

Here, $u_1 = 1, u_2 = -2$ and $\varepsilon_{12} = (-1)(1)(-2) = 2$.

By the Laurent Phenomenon (Theorem 7.1.1), we get Laurent sequences corresponding to every period 1 quiver. For $u_1, u_2, \ldots, u_r \in \mathbb{Z}$, set $u_{n-t} = u_t$ for $t = 1, \ldots, r-1$. The corresponding sequence is given by the recurrence:

$$x_k x_{k+n} = \prod_{t=1}^{n-1} x_{k+t}^{[u_t]_+} + \prod_{t=1}^{n-1} x_{k+t}^{-[u_t]_-}. \tag{7.1}$$

If $n = 4$, this takes the form:

$$x_k x_{k+4} = x_{k+1}^a x_{k+3}^a + x_{k+2}^b, \qquad \text{for } a, b \in \mathbb{Z}_{\geq 0}.$$

The Somos-4 sequence, discussed in Section 7.1, is the case $a = 1, b = 2$.

7.3 Periodicity in the coefficient case

The two-term version of the Gale-Robinson recurrence (without parameters):

$$x_k x_{k+n} = x_{k+i} x_{k+n-i} + x_{k+j} x_{k+n-j},$$

for positive integers i, j such that $i < j \leq k/2$ is also a special case of equation (7.1). The argument above gives an alternative proof of the following result (in the case where the parameters are equal to 1):

Theorem 7.3.1. *[68, Thm. 1.8] Fix positive integers i, j, k such that $i < j \leq k/2$. Then the nth term of the two-term Gale-Robinson recurrence:*

$$x_k x_{k+n} = \alpha x_{k+i} x_{k+n-i} + \beta x_{k+j} x_{k+n-j}$$

is Laurent in the first k terms, with coefficients in $\mathbb{Z}[\alpha, \beta]$.

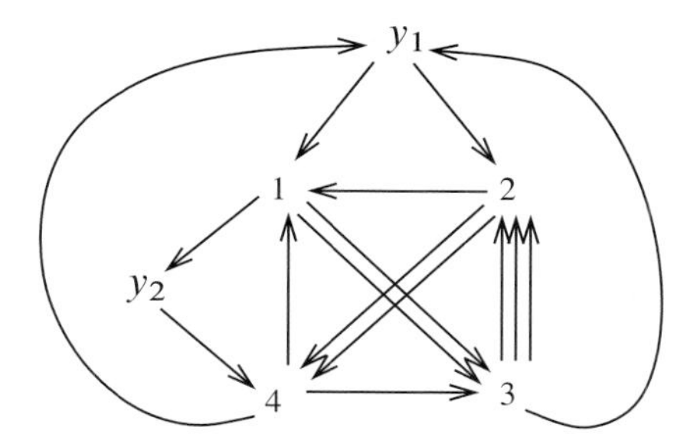

Figure 7.2. A quiver with frozen vertices for the Somos-4 sequence with parameters.

See the comments after Corollary 7.1.2 concerning the proof in [68] and related work [155]. See also, for example, [48] for related work.

Periodicity for cluster algebras with coefficients is also studied in [75], to model recurrences with parameters, such as the above. A quiver $\widetilde{Q}$ with vertices $[1, m]$, with $[n+1, m]$ frozen, is said to be a period 1 quiver if:

$$\mu_1(\widetilde{Q}) = \tilde{\rho}\widetilde{Q},$$

where $\tilde{\rho}$ is the permutation $(1\ 2\ \cdots\ n)$ in the symmetric group of degree m (i.e. fixing $n+1, \ldots, m$). Thus the principal part Q of $\widetilde{Q}$, the full subquiver on vertices I, must be a period 1 quiver in the sense of Section 7.2. The period 1 property puts a restriction on the arrows incident with the frozen vertices. There are arrows from vertex 1 to frozen vertices and also from frozen vertices to vertex 1 in $\widetilde{Q}$ if and only the corresponding recurrence has parameters on both terms. We have the following:

Proposition 7.3.2. *[75, Prop. 10.4] The only binomial recurrences corresponding to period 1 quivers that, when parameters on both monomials are allowed, correspond to period 1 quivers with coefficients, are the two-term Gale-Robinson recurrences.*

Note that this gives, in particular, an alternative perspective on the proof of Theorem 7.3.1.

Example 7.3.3. In Figure 7.2 a quiver with frozen vertices for the Somos-4 sequence with parameters is represented. The corresponding recurrence is:

$$x_k x_{k+4} = y_1 x_{k+1} x_{k+3} + y_2 x_{k+2}^2$$

Adding more frozen vertices to the quiver would give the same recurrence, but with monomials in the coefficients (the frozen variables), with disjoint support, replacing y_1 and y_2 (so essentially gives the same recurrence, defining $\tilde{y_1}$ and $\tilde{y_2}$ to be these two monomials).

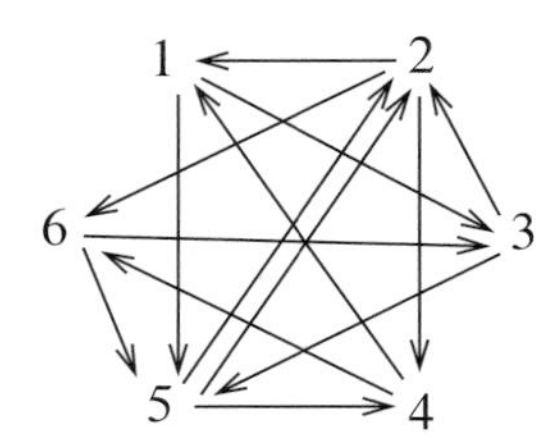

Figure 7.3. Quiver associated to a Del Pezzo surface of degree 3.

7.4 Higher period quivers

In general, the sequences obtained are interesting from a discrete integrable system perspective and are studied that way in [73], [74], [75], [93].

Remark 7.4.1. Periodic quivers arise in the context of supersymmetric quiver gauge theories. E.g. S_4 is associated to a del Pezzo 1 surface; see [54, §4] and [75, §11].

Definition 7.4.2. [75, §2] A quiver Q (without frozen vertices) on vertices $\{1, \ldots, n\}$ is said to be of *period a* if

$$\mu_a \ldots \mu_1 Q = \rho^a Q.$$

For example the quiver in Figure 7.3, associated to a del Pezzo 3 surface, has period 2; see again [54, §4] and [75, §11].

A particularly interesting case is $a = n$, i.e. a quiver has period n if

$$\mu_n \mu_{n-1} \ldots \mu_1 Q = \rho^n Q = Q.$$

If Q is acyclic and the vertices numbered so that the arrow points towards the smaller vertex, we have seen (Lemmas 5.5.2 and 5.5.3) that it has period n: in fact all of the mutations take place at sinks.

Any quiver of period dividing n will also have period n, so all of the above quivers are also examples of period n quivers. In fact the primitives are acyclic, but this is not true for a general period 1 quiver (e.g. S_4). Quivers of higher period are more difficult to classify. Some partial results are available in [75].

7.5 Categorical periodicity

B. Keller has used periodicity of quiver mutations to good effect in the article [103] and we now summarize this.

Let Γ, Γ' be Dynkin diagrams with vertex sets I, I' and Cartan matrices C, C'. Let

$$A = 2Id_{|I|} - C, \quad A' = 2Id_{|I'|} - C',$$

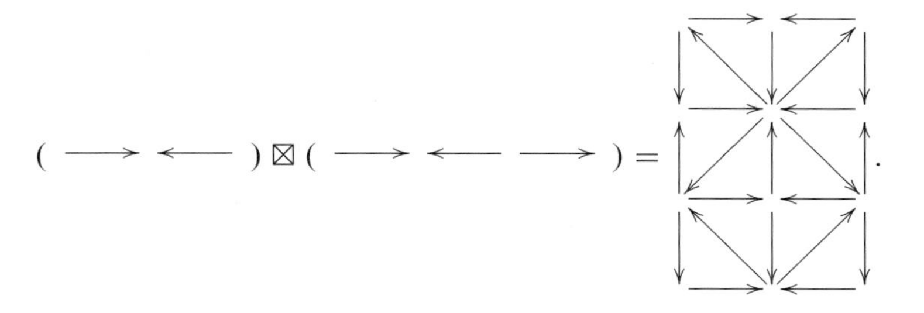

Figure 7.4. Example of a triangle tensor product of alternating quivers.

where Id_m represents the $m \times m$ identity matrix. Let h, h' be the Coxeter numbers of Γ, Γ'. Then the *Y-system of algebraic equations* associated to the pair (Γ, Γ') is a collection of countably many recurrences in variables $Y_{i,i',t}$ where $i \in I, i' \in I', t \in \mathbb{Z}$:

$$Y_{i,i',t-1}Y_{i,i',t+1} = \frac{\prod_{j \in I}(1 + Y_{j,i',t})^{a_{ij}}}{\prod_{j' \in I'}(1 + Y_{i,j',t}^{-1})^{a'_{i'j'}}}.$$

The periodicity conjecture was first formulated by Al. B. Zamolodchikov [165] in the context of the thermodynamic Bethe Ansatz.

Theorem 7.5.1. *(Periodicity conjecture) [103, Thm. 2.3] All solutions to the above system of equations are periodic in t with period dividing $2(h + h')$.*

The proof uses the generalized cluster category, which was introduced in [2], generalising earlier definitions in [23] (the acyclic case) and [27] (type A_n), to get a categorical periodicity corresponding to the Y-system periodicity.

We restrict to the simply-laced case here for simplicity. Let Q, Q' be alternating orientations of Γ, Γ'. Then the *tensor product* of Q and Q' has vertices $Q_0 \times Q'_0$, where Q_0 (resp. Q'_0) is the set of vertices of Q (resp. Q').

The number of arrows from (i, j) to (i', j') is given by:

- the number of arrows from j to j' if $i = i'$;
- the number of arrows from i to i' if $j = j'$;
- zero, otherwise.

The *triangle tensor product* $Q \boxtimes Q'$ is obtained from $Q \otimes Q'$ by adding rr' arrows from (j, j') to (i, i') whenever Q has r arrows from i to j and Q' has r' arrows from i' to j'.

Example 7.5.2. For example, we have the triangle tensor product shown in Figure 7.4.

For a vertex i of Q, let $\varepsilon(i) = +$ (respectively, $-$) if i is a source (respectively, a sink) in Q. For $\sigma, \sigma' \in \{+, -\}$, let

$$\mu_{\sigma,\sigma'} = \prod_{\varepsilon(i)=\sigma,\ \varepsilon(i')=\sigma'} \mu_{(i,i')},$$

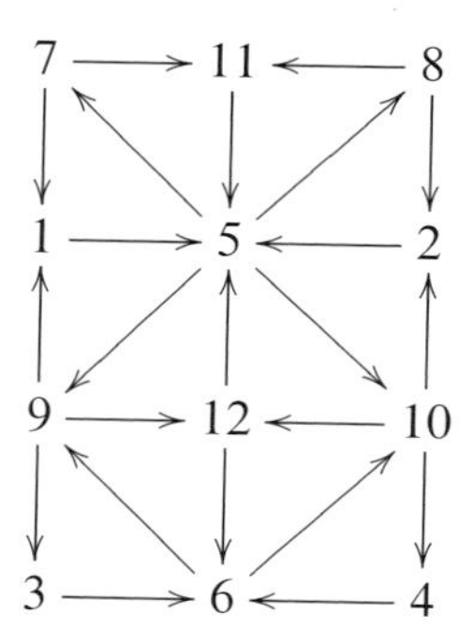

Figure 7.5. This numbering makes the quiver in Figure 7.4 12-periodic.

a product of mutations of $Q \boxtimes Q'$, with the product taken over vertices (i, i') of $Q \boxtimes Q'$. It is easy to check that, since sources (respectively, sinks) of Q and Q' are always separated by a vertex, all the mutations in the product commute with each other, so the order in which they are composed does not matter.

Then we have:

Lemma 7.5.3. *[103, §3] Let Q, Q' be alternating orientations of the simply-laced Dynkin diagrams Γ, Γ'. Then*

$$\mu_{-,+}\mu_{+,+}\mu_{-,-}\mu_{+,-}(Q \boxtimes Q') = Q \boxtimes Q'.$$

Note that each vertex of $Q \boxtimes Q'$ appears exactly once in the sequence of mutations in the lemma, so if the vertices are numbered with those appearing in $\mu_{+,-}$ first, then those in $\mu_{-,-}$, $\mu_{+,+}$ and finally those in $\mu_{-,+}$, we see that $Q \boxtimes Q'$, with this numbering, is $|I||I'|$-periodic in the sense of Definition 7.4.2. Note that $|I||I'|$ is the number of vertices of $Q \boxtimes Q'$. For the example above, the numbering is shown in Figure 7.5.

S. Fomin and A. Zelevinsky, in [72, Defn. 2.9], consider the notion of a Y-pattern associated with a quiver Q with no loops or oriented cycles on vertices $1, \dots, n$ and a free generating set $\mathbf{Y} = \{Y_1, \dots, Y_n\}$ of $\mathbb{Q}(y_1, \dots, y_n)$, the rational function field in n indeterminates over $\mathbb{Q}$. The pair $(Q, \mathbf{Y})$ is called a Y-seed, and is mutated in a manner analogous to cluster mutation. The quiver mutates in the same way as Definition 2.3.2, while the mutation of Y at k is the tuple $\mathbf{Y}' = \{Y_1', Y_2', \dots, Y_n'\}$ where

$$Y_i' = \begin{cases} Y_i^{-1} & \text{if } i = k; \\ Y_i(1 + Y_k^{-1})^{-a} & \text{if there are } a \geq 0 \text{ arrows } k \to i; \\ Y_i(1 + Y_k)^a & \text{if there are } a \geq 0 \text{ arrows } i \to k. \end{cases}$$

If Q has period n, the *restricted Y-pattern* is obtained by only mutating (Q, Y) by applying $\mu_n \dots \mu_1$ or its inverse. A key step in the proof is then the translation of the conjecture to the following proposition.

Proposition 7.5.4. *[103, Lem. 3.7] The periodicity conjecture holds for the pair* (Γ, Γ') *of Dynkin diagrams if and only if the restricted* Y*-pattern as defined above associated to* $Q \boxtimes Q'$ *is periodic of period dividing* $h + h'$.

The property in Proposition 7.5.4 is then proved using the categorical interpretation, i.e. the generalized cluster category [2] mentioned above.

8 Quivers of finite mutation type

We have already considered the classification of cluster algebras of finite type. A different question is which quivers have finite mutation type — i.e. are mutation equivalent to only finitely many other quivers (ignoring the cluster variables). There is a very beautiful answer [51] in terms of the quivers associated to marked Riemann surfaces in [65].

8.1 Classification

A quiver Q is of *finite mutation type* if there are finitely many quivers mutation equivalent to it. Thus, any quiver in a seed of a cluster algebra of finite type has finite mutation type, but the converse is not true.

Theorem 8.1.1. *[51] The quivers of finite mutation type are as follows:*

(a) Quivers with 2 vertices and r arrows from one to the other, where $r \geq 0$.

(b) A quiver arising from a triangulation of a marked surface with boundary (as in [65]).

(c) A quiver mutation equivalent to an orientation of E_6, E_7 or E_8.

(d) A quiver mutation-equivalent to one of those in Figure 8.1.

Remark 8.1.2. The quivers X_6 and X_7 in Figure 8.1 first appeared in [41, Prop. 4]. The skew-symmetrizable matrices of finite mutation type are classified in [52, Thm. 5.13] using an extension of the work in [65] for surfaces to the orbifold setting in [53] and folding arguments (in particular, a notion of unfolding introduced by A. Zelevinsky). We also note that the cluster algebras corresponding to the quivers in (b) have a beautiful interpretation in hyperbolic geometry, in terms of a appropriate decorated Teichmüller space of the surface [66] (see also references therein).

The quivers in case (b) are defined as follows. Fix a connected, oriented Riemann surface S with boundary together with a finite set M of marked points in the closure of S; thus points in M are either interior or on the boundary of S. Interior points of M are referred to as *punctures*. Assume that M is non-empty, that there is at least one marked point on each boundary component, and that (S, M) is not one of:

(a) a sphere with 1, 2 or 3 interior points;

(b) a monogon with 0 or 1 interior points;

(c) a digon or a triangle with no interior points.

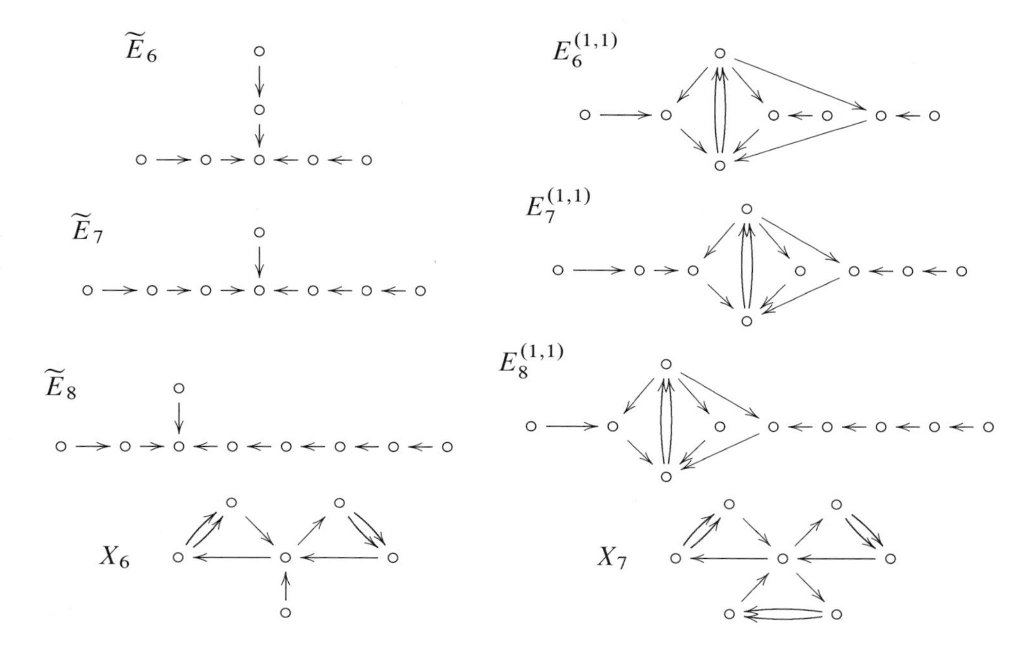

Figure 8.1. Part (d) of Theorem 8.1.1.

Consider simple arcs in (S, M) up to isotopy, with the restriction that the end points lie in M, no other points lie on the boundary, and that the arc is not contractible onto the boundary. An *ideal triangulation* of (S, M) is a maximal set of such arcs which do not cross.

Lemma 8.1.3. *Let g be the genus of S, b the number of boundary components of S, p the number of punctures of (S, M) and c the number of boundary marked points. Then the number of arcs in any ideal triangulation of (S, M) is equal to:*

$$n = 6g + 3b + 3p + c - 6.$$

Proof. See, for example [56, §2]. See also [65, Prop. 2.10]. □

Definition 8.1.4. [65, §4] Let $\mathcal{T}$ be an ideal triangulation of (S, M). Then $\mathcal{T}$ gives rise to a quiver, $Q_{\mathcal{T}}$, with vertices corresponding to the arcs in $\mathcal{T}$, as follows.

For $1 \leq i \leq n$, set

$$\pi_{\mathcal{T}}(i) = \begin{cases} j & \text{if there is a self-folded triangle in } \mathcal{T} \text{ folded along } i \text{ and } j \text{ is the enclosing loop — see Figure 8.2;} \\ i & \text{otherwise.} \end{cases}$$

For each triangle T in $\mathcal{T}$ which is not self-folded, let Q_T be the quiver on vertices $1, \ldots, n$ with an arrow from i to j whenever $\pi_{\mathcal{T}}(i), \pi_{\mathcal{T}}(j)$ lie on the boundary of

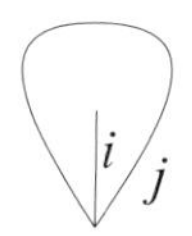

Figure 8.2. Self-folded triangle.

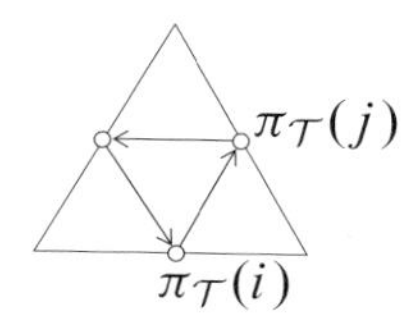

Figure 8.3. Quiver arising from a triangulation.

T and $\pi_{\mathcal{T}}(j)$ follows $\pi_{\mathcal{T}}(i)$ in the order induced by the anticlockwise orientation on the boundary. See Figure 8.3. Note that this is opposite to the convention used in [65, §4], but is chosen to ensure compatibility with the conventions in Chapter 9; see Section 9.5.

Remark 8.1.5. In the situation of a self-folded triangle as in Figure 8.2, $\pi_{\mathcal{T}}(i) = \pi_{\mathcal{T}}(j) = j$, so if there is an arrow $k \to j$ (respectively, $j \to k$) in Q_T for some k and triangle T there is also an arrow $k \to i$ (respectively, $i \to k$) in Q_T.

Then set

$$Q_{\mathcal{T}} = \sum_{\substack{T \text{ triangle in } \mathcal{T} \\ T \text{ not self-folded}}} Q_T.$$

For some examples, see Figures 8.4 and 8.5.

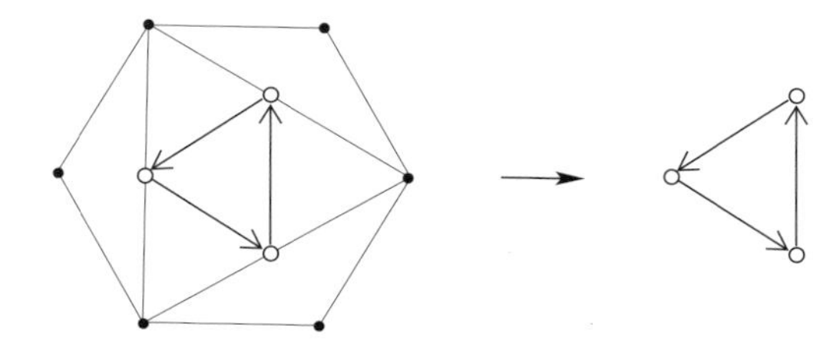

Figure 8.4. A quiver of mutation type A_3 arising from a disk with 6 marked points.

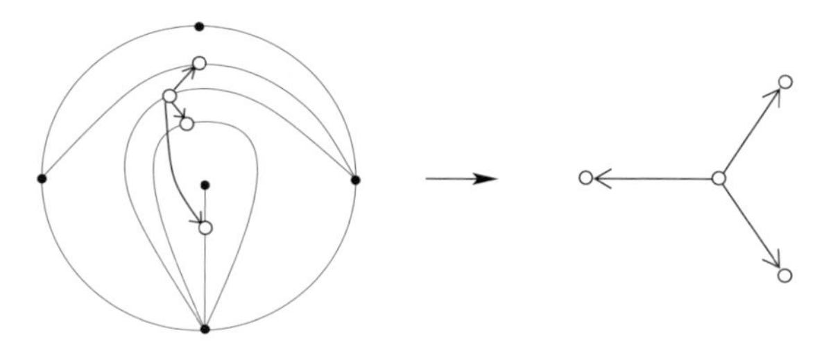

Figure 8.5. A quiver of type D_4 arising from a punctured disk with 4 marked points on its boundary.

8.2 Tagged triangulations

In fact, all of the quivers in the mutation class of a quiver arising from a triangulation of a surface in this way can be obtained from some triangulation of the surface. But for the complete picture (in order to allow arbitrary mutations) we need a more general notion of triangulation, known as a *tagged triangulation* [65, §7].

A *tagged arc* is an arc as above, with the additional possibility that each end of the arc can be tagged. The following conditions must be satisfied:

(a) The arc does not bound a monogon with a single puncture.

(b) Only end-points incident with punctures can be tagged.

(c) For a loop, either both ends are tagged or both ends are untagged.

If γ is a tagged arc, we denote the arc obtained from γ by removing its tags by $|\gamma|$.

Two tagged arcs γ, γ' are said to be *compatible* if the following conditions are satisfied:

(a) The arcs $|\gamma|$ and $|\gamma'|$ do not cross.

(b) If $|\gamma| \neq |\gamma'|$ and γ and γ' have a common endpoint p, then the ends of γ and γ' must either both be untagged or both tagged.

(c) If $|\gamma| = |\gamma'|$ then at least one end of γ must be tagged in the same way as the corresponding end of γ'.

Note that, in (c), if $|\gamma| = |\gamma'|$ is a loop, then both ends must be tagged the same way in γ, and also in γ', so γ and γ' can only be compatible if they are equal.

A tagged triangulation of (S, M) is a maximal collection of compatible tagged arcs in (S, M). The quiver of a tagged triangulation $\mathcal{T}$ is defined in the same way as for an ordinary triangulation, equation (8.1), ignoring the tags on the arcs. Recall that a surface is *closed* if and only if it has no boundary components.

Theorem 8.2.1. *[65, 66] Let $\mathcal{T}$ be a tagged triangulation of (S, M). Then:*

(a) The number of tagged arcs in $\mathcal{T}$ is the same as the number of arcs in a triangulation of (S, M), i.e.

$$n = 6g + 3b + 3p + c - 6.$$

(b) *Let $\mathcal{T}$ be a tagged triangulation containing an arc γ. Then $\mathcal{T}\setminus\{\gamma\}$ is contained in exactly two tagged triangulations, $\mathcal{T}$ and $\mathcal{T}'$. The tagged triangulation $\mathcal{T}'$ is said to be obtained from $\mathcal{T}$ by a* flip.

(c) *In the situation in (b), $Q_{\mathcal{T}'}$ can be obtained from $Q_{\mathcal{T}}$ by quiver mutation at the vertex corresponding to γ.*

(d) *The cluster variables of the cluster algebra $\mathcal{A}$ associated to the quiver of a tagged triangulation $\mathcal{T}$ of (S, M) are in bijection with the tagged arcs in (S, M) appearing in tagged triangulations reachable from $\mathcal{T}$ by flips. This induces a bijection between clusters of $\mathcal{A}$ and such reachable tagged triangulations of (S, M). The flip of tagged triangulations corresponds to cluster mutation.*

Proof. For (a) and (b), see [65, Thm. 7.9]. For (c), see [65, Lemma 9.7]. For (d), see [65, Thm. 7.11] and [66, Thm. 6.1]. □

We also mention a result from [65], which says that in Theorem 8.2.1(d), every tagged triangulation is reachable except in some special cases. Let $E(S, M)$ be the graph with vertices given by the tagged triangulations and edges given by flips.

Recall that a surface is closed if has no boundary.

Proposition 8.2.2. *[65, Prop. 7.10] The graph $E(S, M)$ is connected if and only if (S, M) is not a closed surface with exactly one puncture. If (S, M) is a closed surface with exactly one puncture then $E(S, M)$ has two connected components, each isomorphic to the other: one in which the ends of every arc are untagged, and one in which they are all tagged.*

We next give some examples, based on a table in [65, §2].

Example 8.2.3. (a) [67, §1], [71, §3.5] A disk with $n + 3$ marked points on the boundary has a triangulation giving rise to a quiver of type A_n. See Figure 8.6 for the case $n = 5$. In general, the triangulation of the disk with all arcs incident with a single marked point on the boundary gives rise to the linearly oriented quiver of type A_n and the zig-zag triangulation illustrated in Figure 6.7 gives rise to an alternating orientation of type A_n. This model of type A_n corresponds to the associahedron model discussed in Section 6.6.

In this case, none of the arcs can be tagged (as there are no punctures, and tags can only be at the ends of arcs incident with punctures), so clusters correspond to triangulations of the disk, which can be regarded as triangulations of a regular $(n + 3)$-sided polygon.

(b) [65, Ex. 2.12] (see also [150]) A disk with n marked points on the boundary and a single puncture has a triangulation giving rise to a quiver of type D_n. See Figure 8.5 for an example of type D_4.

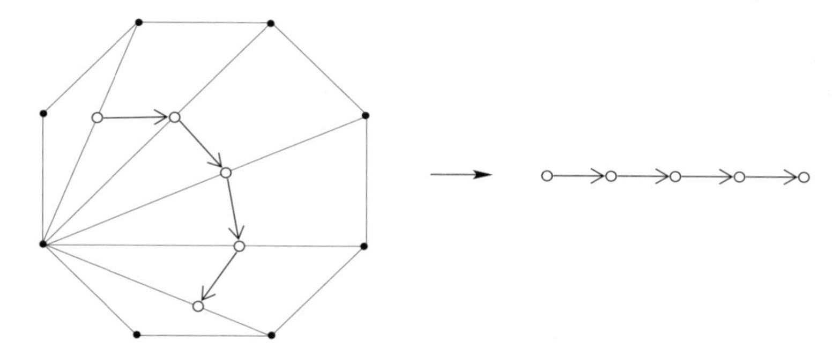

Figure 8.6. Triangulation of a disk with 8 marked points on its boundary.

(c) [65, Ex. 2.12] An annulus with r marked points on one boundary component and s marked points on the other has a triangulation giving rise to a quiver of type $\tilde{A}_{r,s}$, i.e. a quiver whose underlying graph is an $(r+s)$-cycle, with r arrows in one direction and s arrows in the other direction.

(d) [65, Ex. 2.12] A disk with $n-2$ marked points on its boundary and two punctures has a triangulation giving rise to a quiver of type $\widetilde{D}_n$.

We remark that a generalized version of a cluster algebra, known as a quasi-cluster algebra, has recently been associated to a non-orientable surface in [46]. We also note that the surface model has been used to give explicit expressions for cluster variables in the cluster algebra associated to a surface. For the unpunctured case this was done using certain paths in the surface in [152], with a general coefficient version being given in [151]. An approach using perfect matchings (again for the unpunctured case) was given in [131] (see also [130]). This was then generalized to the punctured case in [132] and further developed in [29].

9 Grassmannians

A result of J. S. Scott [153] states that the homogeneous coordinate ring of the Grassmannian of k-subspaces of an n-dimensional space can be regarded as a cluster algebra. The combinatorics of this result is very beautiful, involving certain planar diagrams labelled by minors, introduced by A. Postnikov [143]. For an alternative proof, we refer to [83] and references therein.

9.1 Exterior powers

Let V denote the complex vector space $\mathbb{C}^n$. Recall that $T(V)$ denotes the tensor algebra:

$$T(V) = \mathbb{C} \oplus V \oplus V \oplus V \oplus \cdots .$$

of V. The *exterior algebra* of V is the quotient

$$\bigwedge(V) = T(V)/J,$$

where J denotes the ideal of $T(V)$ generated by the elements $x \otimes x$ for $x \in V$. We denote the product in $\bigwedge(V)$ by $(x, y) \mapsto x \wedge y$ (recall that this comes from the tensor product in $T(V)$).

Lemma 9.1.1. *Let $x, y \in V$. Then $x \wedge y = -y \wedge x$, i.e. the product in $\bigwedge(V)$ is* anticommutative *on elements of V.*

Proof. First, let $x, y \in V$. Then

$$\begin{aligned} 0 &= (x + y) \wedge (x + y) \\ &= x \wedge x + x \wedge y + y \wedge x + y \wedge y \\ &= x \wedge y + y \wedge x, \end{aligned}$$

and the result follows. □

The kth *exterior power*, $\bigwedge^k(V)$ of V is the subspace of $\bigwedge(V)$ spanned by products of the form $v_1 \wedge v_2 \wedge \cdots \wedge v_k$, with $v_r \in V$ for $1 \leq r \leq k$. We have:

$$\bigwedge(V) = \bigoplus_{k=0}^{\infty} \bigwedge^k(V).$$

As usual, $e_1, e_2, \ldots, e_n$ denotes the natural basis of V, with $(e_i)_j = \delta_{ij}$ for all i, j. Then the vectors

$$e_{i_1} \wedge e_{i_2} \wedge \cdots \wedge e_{i_k},$$

with $1 \leq i_1 < i_2 < \cdots < i_k \leq n$ form a basis for $\bigwedge^k(V)$. For an element $x \in \bigwedge^k(V)$, we write the coefficient of x when expanded in terms of this basis by $p_{i_1,i_2,\ldots,i_k}(x)$. Thus $p_{i_1,i_2,\ldots,i_k}$ is a linear map from $\bigwedge^k(V)$ to $\mathbb{C}$.

An element x of $\bigwedge^k(V)$ is said to be *decomposable* if if it is of the form $x = v_1 \wedge v_2 \wedge \cdots \wedge v_k$ for some k, where $v_1, v_2, \ldots, v_k$ form a linearly independent set of vectors in V (note that the wedge product is zero if the vectors are not linearly independent). Given such a linearly independent set, $v_1, v_2, \ldots, v_k$, we can form a $k \times n$ matrix M of rank k whose i, j entry is $M_{ij} = (v_i)_j$, i.e. the jth entry of v_i, so that the rows are the vectors v_i.

We denote the minor of M corresponding to rows $a_1, a_2, \ldots, a_r$ and columns $b_1, b_2, \ldots, b_r$ by $\Delta^{a_1,a_2,\ldots,a_r}_{b_1,b_2,\ldots,b_r}(M)$. The following is easy to check, using the expansion of the v_i in terms of $e_1, e_2, \ldots, e_n$.

Lemma 9.1.2. *Let $x = v_1 \wedge v_2 \wedge \cdots \wedge v_k$ be a decomposable element of $\bigwedge^k(V)$. Then $p_{i_1,i_2,\ldots,i_k}(x)$ is the minor $\Delta^{1,2,\ldots,k}_{i_1,i_2,\ldots,i_k}(M)$, where M is the matrix defined above.*

9.2 The Grassmannian

The *Grassmannian* $Gr(k,n)$ is the set of k-dimensional subspaces of V. If U is such a subspace and $v_1, v_2, \ldots, v_k$ is a basis of U, consider the decomposable element

$$\omega = v_1 \wedge v_2 \wedge \cdots \wedge v_k \in \bigwedge^k(V).$$

This element is non-zero and does not depend upon the choice of basis, up to a non-zero scalar. It follows that the tuple $(p_{i_1,i_2,\ldots,i_k}(\omega))_{1 \leq i_1 < i_2 < \ldots < i_k \leq n}$ is a well-defined element of the projective space $\mathbb{P}^N$, where $N = \binom{n}{k} - 1$. Thus we obtain a map $\varphi : Gr(k,n) \to \mathbb{P}^N$. The $p_{i_1,i_2,\ldots,i_k}$ are referred to as *Plücker coordinates.*

We extend the definition of $p_{i_1,i_2,\ldots,i_k}$ to arbitrary elements $i_1, i_2, \ldots, i_k$ in I by setting $p_{i_1,i_2,\ldots,i_k} = 0$ if $i_r = i_s$ for some r, s and setting

$$p_{i_1,i_2,\ldots,i_k} = \operatorname{sgn}(\pi) p_{j_1,j_2,\ldots,j_k}$$

if the i_r are distinct,

$$\{i_1, i_2, \ldots i_k\} = \{j_1, j_2, \ldots, j_k\},$$

$j_1 < j_2 < \cdots < j_k$ and π is the permutation:

$$\pi = \begin{pmatrix} i_1 & i_2 & \cdots & i_k \\ j_1 & j_2 & \cdots & j_k \end{pmatrix}.$$

The *Plücker relations* for $Gr(k,n)$ are the relations:

$$\sum_{r=0}^{k} (-1)^r p_{i_1,i_2,\ldots,i_{k-1},j_r} p_{j_0,\ldots,\widehat{j_r},\ldots,j_d} = 0, \tag{9.1}$$

where the sum is taken over all tuples satisfying $1 \leq i_1 < i_2 < \cdots < i_{k-1} \leq n$ and $1 \leq j_0 < j_1 < \cdots < j_k \leq n$ and the hat indicates omission.

Theorem 9.2.1. *(a) An element $x \in \bigwedge^k(V)$ is decomposable if and only if the Plücker relations* (9.1) *hold.*

(b) The image of φ is the subset of $\mathbb{P}^N$ consisting of elements for which the Plücker relations (9.1) *hold.*

(c) The map $\varphi : Gr(k,n) \to \mathbb{P}^N$ is injective (it is known as the Plücker embedding*).*

(d) The image of φ is an irreducible projective variety and hence so is $Gr(k,n)$.

Proof. For (a),(b) and (c), see e.g. [100, §3.4]; or for (b) and (c) see [128, §14]. Note that (b) follows from (a) and the definition of φ. For (d), see e.g. [76, §8,§9] or [11, §12.5.8]. □

Example 9.2.2. For the Grassmannian $Gr(2,4)$ there is a single Plücker relation:

$$p_{12}p_{34} - p_{13}p_{24} + p_{14}p_{23} = 0.$$

See Section 9.3.

We follow [89, §2]. Let $Y \subseteq \mathbb{P}^r$ be a projective variety. Then $R = \mathbb{C}[x_0, x_1, \ldots, x_r]$ is a graded ring, with the weight of each x_i being 1. The homogeneous ideal $J(Y)$ of Y is then the ideal of R generated by the homogeneous elements of R which vanish on Y. The *homogeneous coordinate ring* of Y is the quotient $\mathbb{C}[Y] = R/J(Y)$.

There is a natural projection map $\pi : \mathbb{C}^{r+1} \setminus \{0\} \to \mathbb{P}^r$ taking $(a_0, a_1, \ldots, a_r)$ to the line $[a_0 : a_1 : \ldots : a_r]$. The *affine cone* $C(Y)$ over Y is the preimage of Y under π, together with the zero element. By e.g. [89, Exercise 2.10], $C(Y)$ is an affine variety whose ideal is $J(Y)$, regarded as an ordinary ideal in R without its grading. It follows that the coordinate ring of $C(Y)$ coincides with the homogeneous coordinate ring of Y.

It can be seen from Theorem 9.2.1 that the affine cone of $Gr(k,n)$ can be identified with the decomposable elements of $\bigwedge^k(n)$ (together with zero). Again by Theorem 9.2.1, its coordinate ring, i.e. the homogeneous coordinate ring $\mathbb{C}[Gr(k,n)]$, is the quotient of the polynomial ring in generators $x_{i_1,i_2,\ldots,i_k}$ for $1 \leq i_1 < i_2 < \cdots < i_k \leq n$ by the ideal generated by the Plücker relations.

9.3 The Grassmannian $Gr(2,n)$

We consider the case $k = 2$, i.e. the Grassmannian $Gr(2,n)$. The Plücker relations take the form

$$p_{i_1 j_0} p_{j_1 j_2} - p_{i_1 j_1} p_{j_0 j_2} + p_{i_1 j_2} p_{j_0 j_1} = 0$$

for all $1 \leq i_1 \leq n$ and $1 \leq j_0 < j_1 < j_2 \leq n$. If $i_1 < j_0$, this can be written as:

$$p_{ab}p_{cd} - p_{ac}p_{bd} + p_{ad}p_{bc} \tag{9.2}$$

for all $1 \leq a < b < c < d \leq n$. If $i_1 = j_0$, it reduces to $-p_{i_1j_1}p_{i_1j_2} + p_{i_1j_2}p_{i_1j_1} = 0$, which is automatically satisfied (as the homogeneous coordinate ring is commutative). If $j_0 < i_1 < j_2$, it becomes

$$-p_{j_0i_1}p_{j_1j_2} - p_{i_1j_1}p_{j_0j_2} + p_{i_1j_2}p_{j_0j_1},$$

which is the negative of:

$$p_{j_0i_1}p_{j_1j_2} - p_{j_0j_1}p_{i_1j_2} + p_{j_0j_2}p_{i_1j_1},$$

which is an instance of (9.2) with $a = j_0$, $b = i_1$, $c = j_1$, $d = j_2$. Similarly, the other cases reduce to trivial relations or instances of (9.2). By Theorem 9.2.1, we have the following:

Lemma 9.3.1. *The homogeneous coordinate ring of $Gr(2,n)$ is the quotient of the polynomial ring in variables p_{ab} for $1 \leq a < b \leq n$ subject to the relations*

$$p_{ab}p_{cd} - p_{ac}p_{bd} + p_{ad}p_{bc} = 0 \tag{9.3}$$

for $1 \leq a < b < c < d \leq n$.

Note that the Plücker coordinates p_{ab} can be parametrized by the diagonals and edges of a regular polygon P_n with vertices $1, 2, \ldots, n$ labelled clockwise around the boundary. The coordinate p_{ab} (with $a < b$) corresponds to the diagonal or boundary edge joining a and b. The relation (9.3) can then be interpreted as stating that in any quadrilateral with boundary vertices a, b, c, d with $1 \leq a < b < c < d \leq n$, the product of the Plücker coordinates corresponding to the two diagonals is equal to the sum of the products of the pairs of coordinates corresponding to pairs of opposite sides of the quadrilateral. See Figure 9.1.

Definition 8.1.4 can be modified slightly to produce a new quiver $\widetilde{Q}_{\mathcal{T}}$ from a triangulation $\mathcal{T}$ of a surface. The vertices of $\widetilde{Q}_{\mathcal{T}}$ are the arcs in $\mathcal{T}$ together with the boundary edges (regarded as frozen vertices), and the arrows are defined in the same way as in Definition 8.1.4, except that the frozen vertices are now included, and there are no arrows between frozen vertices. See [55, 84] and [66, Rk. 5.16].

We can now state:

Theorem 9.3.2. *[67, §1] Let $n \geq 5$ be an integer. Then the homogeneous coordinate ring $\mathbb{C}[Gr(2,n)]$ is a cluster algebra over $\mathbb{C}$ with cluster variables given by the p_{ab}, ab a diagonal in P_n, and coefficients given by the remaining Plücker coordinates, i.e.*

$$p_{12}, p_{23}, \ldots, p_{n-1,n}, p_{1,n},$$

corresponding to the boundary edges of P_n. The seeds are in bijection with the triangulations of P_n, with the elements of the cluster of the seed being the Plücker

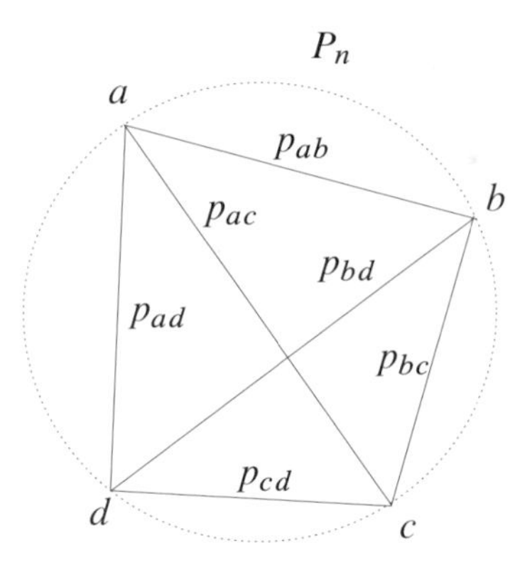

Figure 9.1. The Plücker relation $p_{ac}p_{bd} = p_{ab}p_{cd} + p_{ad}p_{bc}$ in $\mathbb{C}[Gr(2,n)]$.

coordinates corresponding to the edges in the triangulation, and the quiver of the seed being $\widetilde{Q}_{\mathcal{T}}$, where $\mathcal{T}$ is the triangulation. Cluster mutation corresponds to the quadrilateral flip of a triangulation.

Note that here the coefficients are not inverted. We refer to the comment after Theorem 9.4.3 for more about this.

Compare this example with Example 8.2.3. We see that $\mathbb{C}[Gr(2,n)]$ can be regarded as a cluster algebra of type A_{n-3} with n coefficients. See also [144]. The cluster algebra $\mathbb{C}[Gr(2,n)]$ is sometimes referred to as a *Ptolemy cluster algebra* (see [149]) because of the relationship with Ptolemy's Theorem (which states that for a convex quadrilateral inscribed in a circle, the product of the two diagonals coincides with the sum of the products of the pairs of opposite sides).

9.4 The Grassmannian $Gr(k,n)$

For the general case, we need a more general notion of diagram in the plane, generalizing the triangulations for the $Gr(2,n)$ case. We use the notation S_n for the symmetric group of degree n. We follow [153]. (See also [83, §4] and references therein).

Definition 9.4.1. See [143, §14] (and also [153, §3]). Let $n \geq 1$ be an integer, and fix a permutation $\pi \in S_n$ (for simplicity, we consider only the case where π has no fixed points). We consider a disk in the plane with $2n$ vertices equally spaced clockwise around its boundary (although we shall relax the equal spacing requirement in figures), labelled $1', 1, 2', 2, \ldots, n', n$. A *Postnikov diagram*, *Postnikov arrangement* or *alternating strand diagram* of type π, or *π-diagram*, is a collection D of n oriented paths (smooth curves) $\gamma_1, \gamma_2, \ldots, \gamma_n$ in the interior of the disk, with γ_i starting at i and ending at $\pi(i)'$, satisfying the following conditions:

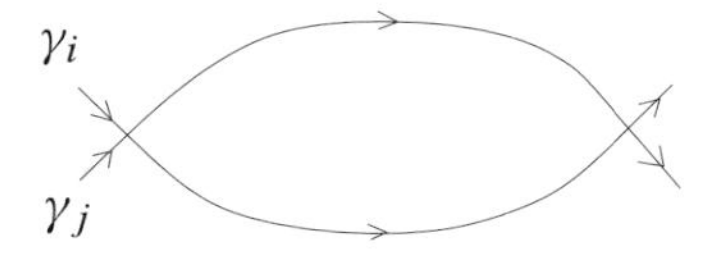

Figure 9.2. Forbidden configuration in a Postnikov diagram.

Figure 9.3. Postnikov diagram equivalence.

(a) Each path γ_i has no self-intersections;

(b) All intersections are transversal, i.e. at most two paths meet at any given point and the tangent vectors to the paths are not parallel at a point of intersection;

(c) Following a path from its start to its finish, the orientations of the other paths crossing it alternate, the first crossing left to right, the next crossing right to left, and so on, with the final one crossing right to left;

(d) The configuration in Figure 9.2 is not allowed (even if other paths cross the region bounded by γ_i and γ_j in the figure).

We consider Postnikov diagrams up to isotopy, and also up to the move illustrated in Figure 9.3 (with no paths allowed to cross the bounded region).

The paths in a Postnikov diagram divide up the disk into regions. We refer to those bounded entirely by paths as *internal* regions and the remaining regions as *boundary* regions. We orient the boundary of the disk clockwise. The boundary of each region consists of a number of oriented line segments. Then it is either the case that the line segments are consistent with an orientation of the boundary of the region (an *oriented region*) or that the line segments alternate in orientation (an *alternating region*).

Each alternating region is labelled with a k-subset K of I (i.e. a subset with k elements) as follows. Each path γ_i divides the disk into two parts, the part to the left of the path when traversing it and the part to the right. Then an element $i \in I$ lies in K if and only if the region lies in the part of the disk to the left of the path γ_i.

We fix the permutation $\pi = \pi_{k,n}$ sending i to $i + k \mod n$ (taking representatives in I). Then the boundary alternating regions are labelled by the n k-subsets of the form $C_1 = [1,k], C_2 = [2,k+1], \dots, C_n = [n,n+k]$ (with the elements again reduced modulo n). We have the following:

Proposition 9.4.2. *(A. Postnikov; see [153, Prop. 5])*

(a) The number of alternating regions in a $\pi_{k,n}$-diagram is $k(n-k)+1$.

(b) Each subset of I labelling an alternating region is a k-subset, i.e. it has k elements.

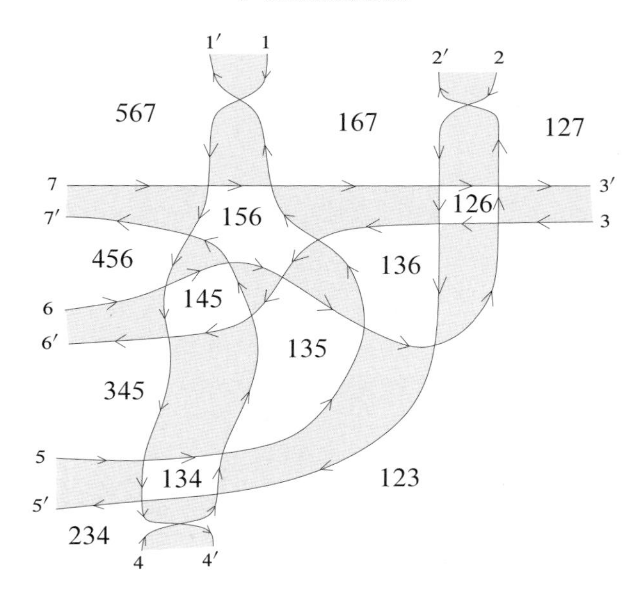

Figure 9.4. An example of a $\pi_{3,7}$-diagram.

(c) *If K is any k-subset of I then there is an alternating region in some $\pi_{k,n}$-diagram labelled with K.*

For an example of a $\pi_{3,7}$-diagram with labelled alternating regions, see Figure 9.4. The oriented regions have been shaded.

Each k-subset can be interpreted as a Plücker coordinate. For a $\pi_{k,n}$-diagram D, we write $\mathbf{p}(D)$ for the set of Plücker coordinates labelling internal alternating regions of D and $\mathbf{c}$ for the set of Plücker coordinates labelling boundary alternating regions of D (listed above); note that this latter set does not depend on the choice of $\pi_{k,n}$-diagram D. We set $\widetilde{\mathbf{p}}(D) = \mathbf{p}(D) \cup \mathbf{c}$; this is the set of labels of all alternating regions of D.

We can also associate a quiver $Q(D)$ to D, with vertices corresponding to the alternating regions. The vertices corresponding to boundary alternating regions are regarded as frozen vertices. The arrows correspond to incidences between alternating regions, with the arrow oriented according to the orientations of the boundaries at the point of incidence; see Figure 9.5. A maximal collection of two-cycles is then cancelled. Note that there is a natural correspondence between the vertices in the quiver $Q(D)$ and the Plücker coordinates in $\widetilde{\mathbf{p}}$.

Fix a $\pi_{k,n}$-diagram D. We define a cluster algebra $\mathcal{A}(D)$ (over $\mathbb{C}$) using the quiver $Q(D)$, with frozen vertices corresponding to the boundary alternating regions, and indeterminates x_K where K varies over the k-subsets labelling all of the alternating regions of D. Note that the indeterminates are in natural bijection with the vertices of the quiver $Q(D)$.

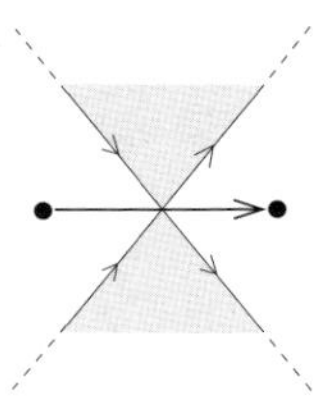

Figure 9.5. Arrow in a quiver associated to a Postnikov diagram.

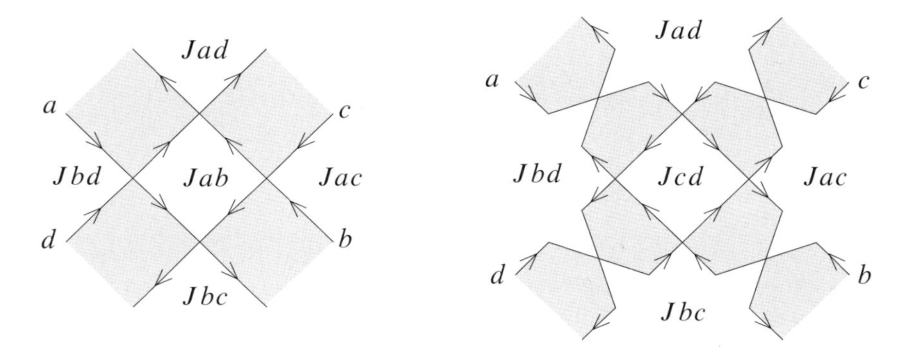

Figure 9.6. Geometric exchange.

Theorem 9.4.3. *[153, Thm. 3] There is an isomorphism $\varphi : \mathcal{A}(D) \xrightarrow{\sim} \mathbb{C}[Gr(k,n)]$. We have $\varphi(x_K) = p_K$ for all k-subsets K labelling an alternating region of D.*

Note that the above holds as stated because in the definition of cluster algebras of geometric type we are using, the coefficients are not inverted (see Definition 2.1.6 and the comment afterwards). If the coefficients are inverted, the corresponding cluster algebra is the coordinate ring of the Zariski open subset of the Grassmannian (which is an affine subvariety) defined by the non-vanishing of the coefficients

$$p_{12}, p_{23}, \dots, p_{n-1,n}, p_{1,n}.$$

See [83, §3]. This subvariety plays a role in the mirror symmetry of the Grassmannian [125, §3.1,7.5].

Given a quadrilateral internal alternating region in a $\pi_{k,n}$-diagram D, a new $\pi_{k,n}$-diagram D' can be constructed by applying the *geometric exchange* illustrated in Figure 9.6. This is a local move — there should be no other paths in the region shown. The alternating regions have been labelled with their subsets: here J denotes a $(k-2)$-subset common to the labels of all the alternating regions in the figure, and Jab denotes the set $J \cup \{a,b\}$, etc.

Then we have:

Theorem 9.4.4. *[153, Thms. 2 and 3]*

(a) *Each $\pi_{k,n}$-diagram D' gives rise to a seed $(\widetilde{\mathbf{x}}(D'), Q(D'))$ in $\mathcal{A}(D)$, with the cluster variables in the cluster of the seed indexed by the k-subsets labelling the alternating regions of D'. We denote the cluster variable corresponding to a k-subset K by x_K.*

(b) *If D'' is obtained from D' by a single geometric exchange, then the mutation of $(\widetilde{\mathbf{x}}(D'), Q(D'))$ at the corresponding cluster variable gives the seed $(\widetilde{\mathbf{x}}(D''), Q(D''))$*

(c) *The isomorphism in Theorem 9.4.3 satisfies $\varphi(x_K) = p_K$ for all k-subsets K of I.*

Thus we see that $\mathbb{C}[Gr(k,n)]$ can be viewed as a cluster algebra, with Postnikov diagrams ($\pi_{k,n}$-diagrams) corresponding to some of its seeds. In Section 9.3 we have seen that $\mathbb{C}[Gr(2,n)]$ has cluster type A_{n-3}, so is of finite type [67, §1], [70, §12.2]. J. S. Scott also proved the result:

Theorem 9.4.5. *[153] The homogeneous coordinate ring $\mathbb{C}[Gr(k,n)]$, with $2 < k \leq n/2$, has finite cluster type if and only if it is $Gr(3,6)$, $Gr(3,7)$ or $Gr(3,8)$. These have finite cluster types D_4, E_6 and E_8, respectively.*

The q-analogues of these coordinate rings have been studied as quantum cluster algebras in [86]. For more information on the Postnikov diagrams discussed above, see [139, 143]. We note also that the cluster algebra structure of the Grassmannian plays a role in the study [111] of soliton solutions of the KP-equation. We also remark that it was recently shown in [62] that the ring of $SL_3(\mathbb{C})$-invariants of tuples of a vectors in $\mathbb{C}^3$ and b linear forms on $\mathbb{C}^3$ (for fixed a, b) also possesses cluster structures; this is closely related to the case $Gr(3,n)$.

9.5 The Grassmannian $Gr(2,n)$ revisited

J. S. Scott also showed that the general $Gr(k,n)$ cluster structure discussed in the last section is compatible with the $Gr(2,n)$ cluster structure discussed in Section 9.3. We will now explain this.

Fix $n \geq 1$ and let P_n be a regular polygon with n sides and vertices $1, 2, \dots, n$ numbered clockwise around the boundary. Let $\mathcal{T}$ be a triangulation of P_n. We can associate a diagram $D(\mathcal{T})$ to $\mathcal{T}$ in the following way.

Firstly, draw three path segments inside each triangle T in $\mathcal{T}$ corresponding to the boundary edges of T in the following way. Label the vertices of T by V_1, V_2 and V_3 clockwise around the boundary. Given a boundary edge $V_{i-1}V_i$, we draw a corresponding path parallel to this edge from a point on the edge V_iV_{i+1} near V_i to a point on the edge $V_{i-1}V_{i+1}$ near V_{i-1}, oriented towards this second point. Subscripts are interpreted modulo 3. See Figure 9.7.

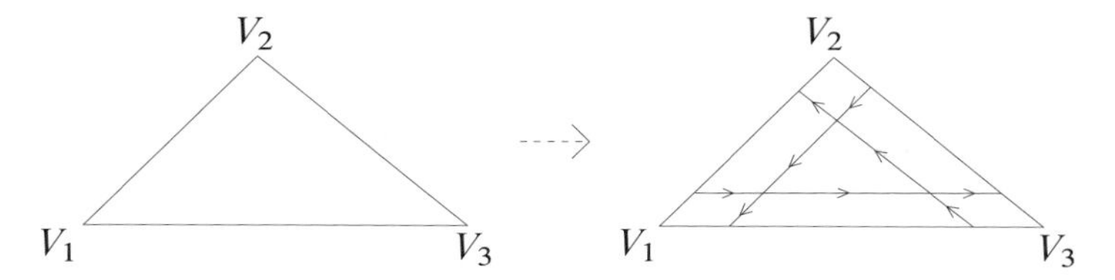

Figure 9.7. The paths in a triangle in $\mathcal{T}$.

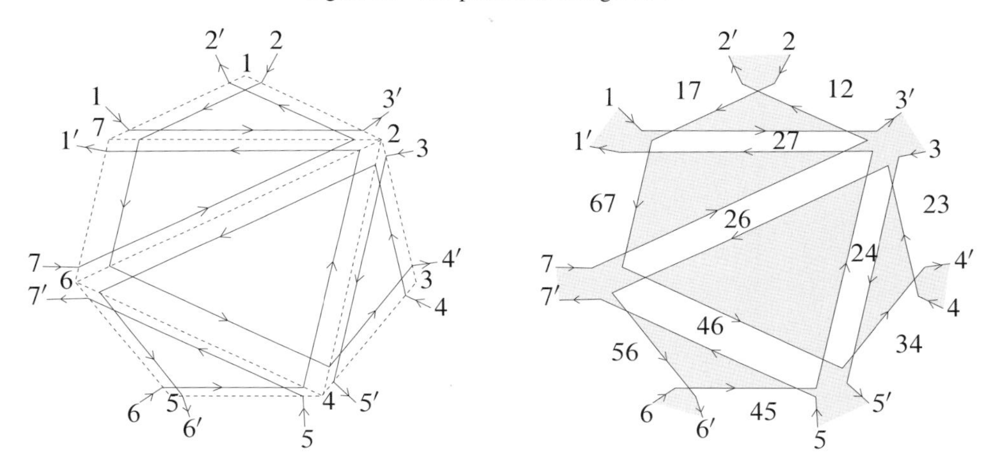

Figure 9.8. The Postnikov diagram associated to a triangulation.

If two triangles in $\mathcal{T}$ are bounded by a common edge, the corresponding paths are matched up along this edge. Paths ending on the boundary of P_n are extended a small amount outside the boundary of P_n. The start of the path immediately clockwise of vertex i of P_n is labelled $\pi_{1,n}(i)$ and the end of the path immediately anticlockwise of vertex i of P_n is labelled $\pi_{1,n}(i)'$. See Figure 9.8 for an example with $n = 7$.

Theorem 9.5.1. *[153, Lem. 1,Cor. 2]*

(a) The map $\mathcal{T} \mapsto D(\mathcal{T})$ gives a bijection between triangulations of P_n and $\pi_{2,n}$-diagrams.

(b) The triangulation $\mathcal{T}'$ can be obtained from $\mathcal{T}$ by a flip if and only if $D(\mathcal{T}')$ can be obtained from $D(\mathcal{T})$ by a geometric exchange.

(c) The seed associated to $\mathcal{T}$ in Theorem 9.3.2 coincides with (the image under φ of) the seed associated to $D(\mathcal{T})$ in Theorem 9.4.4.

The seed corresponding to the triangulation in Figure 9.8 consists of the Plücker coordinates $p_{12}, p_{23}, p_{34}, p_{45}, p_{56}, p_{67}, p_{17}$ (coefficients), the Plücker coordinates $p_{24}, p_{26}, p_{27}, p_{46}$ (cluster variables) and the quiver shown in Figure 9.9.

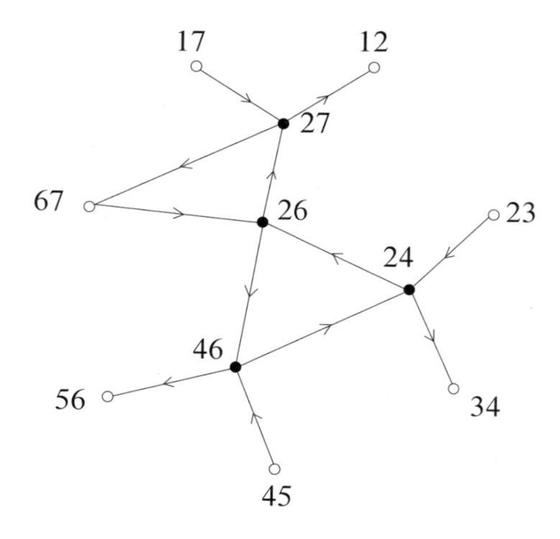

Figure 9.9. The quiver of the triangulation in Figure 9.8.

Bibliography

[1] M. Alim, S. Cecotti, C. Cordova, S. Espahbodi, A. Rastogi and C. Vafa. BPS Quivers and Spectra of Complete N=2 Quantum Field Theories. Preprint arXiv:1109.4941v1 [hep-th], 2011.

[2] C. Amiot. Cluster categories for algebras of global dimension 2 and quivers with potential. *Ann. Inst. Fourier (Grenoble)*, 59 (2009), no. 6, 2525–2590.

[3] C. Amiot. On generalized cluster categories. *In* Representations of Algebras and Related Topics, Proceedings of ICRA XIV, Tokyo, 2010, edited by A. Skowronski and K. Yamagata, Eur. Math. Soc., 2011, 1–53.

[4] I. Assem, M. Blais, T. Brüstle and A. Samson. Mutation classes of skew-symmetric 3×3-matrices. *Comm. Algebra* 36 (2008), no. 4, 1209–1220.

[5] I. Assem and G. Dupont. Friezes and a construction of the Euclidean cluster variables. *J. Pure Appl. Algebra* 215 (2011), no. 10, 2322–2340.

[6] I. Assem, G. Dupont and R. Schiffler. On a category of cluster algebras. Preprint arXiv:1201.5986v1 [math.RT], 2012. J. Pure Appl. Alg., in press.

[7] I. Assem, G. Dupont, R. Schiffler and D. Smith, Friezes, strings and cluster variables. *Glasg. Math. J.* 54 (2012), no. 1, 27–60.

[8] I. Assem, C. Reutenauer and D. Smith. Friezes. *Adv. Math.* 225 (2010), no. 6, 3134–3165.

[9] I. Assem, R. Schiffler and V. Shramchenko. Cluster automorphisms and compatibility of cluster variables. *Proc. Lond. Math. Soc.* (2012) 104 (6): 1271–1302.

[10] I. Assem, D. Simson and A. Skowroński. Elements of the representation theory of associative algebras. Vol. 1. Techniques of representation theory. *London Mathematical Society Student Texts*, 65. Cambridge University Press, Cambridge, 2006.

[11] M. C. Beltrametti, E. Carletti, D. Gallarati and G. Monti Bragadin. Lectures on curves, surfaces and projective varieties. A classical view of algebraic geometry. Translated from the 2003 Italian original by Francis Sullivan. *EMS Textbooks in Mathematics*. European Mathematical Society (EMS), Zürich, 2009.

[12] M. Barot, C. Geiss and A. Zelevinsky. Cluster algebras of finite type and positive symmetrizable matrices. *J. London Math. Soc.* (2) 73 (2006), no. 3, 545–564.

[13] M. Barot and R. J. Marsh. Reflection group presentations arising from cluster algebras. Preprint arXiv:1112.2300v1 [math.GR], 2011. Trans. Amer. Math. Soc., in press.

[14] K. Baur and R. J. Marsh. Frieze patterns for punctured discs. *J. Algebraic Combin.* 30 (2009), no. 3, 349–379.

[15] K. Baur and R. J. Marsh. Categorification of a frieze pattern determinant. *J. Combin. Theory, Ser. A* 119 (2012), 1110–1122.

[16] I. N. Bernštein, I. M. Gel'fand and V. A. Ponomarev, Coxeter functors, and Gabriel's theorem. (Russian) *Uspehi Mat. Nauk* 28 (1973), no. 2 (170), 19–33. English translation in *Russian Math. Surveys* 28 (1973), no. 2, 17–32.

[17] A. Berenstein, S. Fomin, and A. Zelevinsky. Cluster algebras. III. Upper bounds and double Bruhat cells. *Duke Math. J.* 126 (2005), no. 1, 1–52.

[18] D. Berenstein and M. R. Douglas. Seiberg Duality for Quiver Gauge Theories. Preprint arXiv:hep-th/0207027v1, 2002.

[19] F. Bergeron and C. Reutenauer. SL_k-tilings of the plane. *Illinois J. Math.* 54 (2010), no. 1, 263–300.

[20] R. Bott and C. Taubes. On the self-linking of knots. Topology and physics. *J. Math. Phys.* (10) 35 (1994), 5247–5287.

[21] N. Bourbaki. Elements of Mathematics. Lie Groups and Lie Algebras. Chapters 4-6. Springer, 2002. (Originally published as Groupes et Algèbres de Lie, Hermann, Paris, 1968).

[22] A. B. Buan and R. J. Marsh. Cluster-tilting theory. *In* Trends in Representation Theory of Algebras and Related Topics, Workshop August 11-14, 2004, Queretaro, Mexico. Editors J. A. de la Pena and R. Bautista, *Contemporary Mathematics* 406 (2006), 1–30.

[23] A. B. Buan, R. J. Marsh, M. Reineke, I. Reiten, and G. Todorov. Tilting theory and cluster combinatorics. *Adv. Math.* 204, no. 2 (2006), 572–618.

[24] A. B. Buan, R. J. Marsh, I. Reiten, and G. Todorov. Clusters and seeds in acyclic cluster algebras. *Proc. Amer. Math. Soc.* 135, (2007) no. 10, 3049–3060 (electronic). With an appendix coauthored in addition by P. Caldero and B. Keller.

[25] A. Berenstein and A. Zelevinsky. Quantum cluster algebras. *Adv. Math.* 195 (2005), no. 2, 405–455.

[26] R. H. Buchholz and R. L. Rathbun, An Infinite Set of Heron Triangles with Two Rational Medians. *Amer. Math. Monthly* 104 (1997), no. 2, 107–115.

[27] P. Caldero, F. Chapoton and R. Schiffler. Quivers with relations arising from clusters (A_n case). *Trans. Amer. Math. Soc.* 358 (2006), no. 3, 1347–1364.

[28] P. Caldero, F. Chapoton and R. Schiffler. Quivers with relations and cluster tilted algebras. *Algebr. Represent. Theory* 9 (2006), no. 4, 359–376.

[29] I. Canakci and R. Schiffler. Snake graph calculus and cluster algebras from surfaces. *J. Algebra* 382 (2013), 240–281.

[30] R. W. Carter. Cluster algebras. *Textos de Matemática. Série B* [*Texts in Mathematics. Series B*] 37. Universidade de Coimbra, Departamento de Matemática, Coimbra, 2006.

[31] Carter, R. W. Lie algebras of finite and affine type. *Cambridge Studies in Advanced Mathematics*, 96. Cambridge University Press, Cambridge, 2005.

[32] R. W. Carter and R. J. Marsh. Regions of linearity, Lusztig cones and canonical basis elements for the quantized enveloping algebra of type A_4. *J. Algebra*, 234 (2000), no. 2, 545–603.

[33] L. O. Chekhov. Teichmüller theory of bordered surfaces. *SIGMA Symmetry Integrability Geom. Methods Appl.* 3 (2007), Paper 066, 37 pp.

[34] G. Cerulli Irelli, B. Keller, D. Labardini-Fragoso and P.-G. Plamondon. Linear independence of cluster monomials for skew-symmetric cluster algebras. Preprint arXiv:1203.1307v4 [math.RT], 2012. To appear in Compositio Mathematica.

[35] F. Chapoton, S. Fomin, and A. Zelevinsky. Polytopal realizations of generalized associahedra. *Canad. Math. Bull.* 45 (2002), no. 4, 537–566.

[36] P. Caldero and B. Keller. From triangulated categories to cluster algebras. II. *Ann. Sci. Éc. Norm. Supér.* (4), 39 (2006), no. 6, 983–1009.

[37] L. Demonet. Mutations of group species with potentials and their representations. Applications to cluster algebras. Preprint arXiv:1003.5078v2 [math.RT], 2010.

[38] L. Demonet, Laurent Algèbres amassées et algèbres préprojectives: le cas non simplement lacé. (French) [Cluster algebras and preprojective algebras: the non-simply laced case] *C. R. Math. Acad. Sci. Paris* 346 (2008), no. 7–8, 379–384.

[39] L. Demonet, Categorification of skew-symmetrizable cluster algebras. *Algebr. Represent. Theory* 14 (2011), no. 6, 1087–1162.

[40] H. Derksen, J. Weyman and A. Zelevinsky. Quivers with potentials and their representations I: Mutations. *Selecta Math. (N.S.)* 14 (2008), 59–119.

[41] H. Derksen and T. Owen. New graphs of finite mutation type. *Electron. J. Combin.* 15 (2008), no. 1, Research Paper 139, 15 pp.

[42] M. Ding and F. Xu. A quantum analogue of generic bases for affine cluster algebras. Sci. China Math. 55 (2012), no. 10, 2045–2066.

[43] V. Dlab, and C. M. Ringel, Indecomposable representations of graphs and algebras. *Mem. Amer. Math. Soc.* 6 (1976), no. 173.

[44] V. G. Drinfel'd. Hopf algebras and the Yang–Baxter equation. *Soviet Math. Dokl.* 32, 254–258, 1985.

[45] G. Dupont. An approach to non-simply laced cluster algebras. *J. Algebra* 320 (2008), no. 4, 1626–1661.

[46] G. Dupont and F. Palesi. Quasi-cluster algebras from non-orientable surfaces. Preprint arXiv:1105.1560v1 [math.GT], 2011.

[47] G. Dupont and H. Thomas. Atomic bases in cluster algebras of types A and $\widetilde{A}$. Proc. London Math. Soc. (2013), doi:10.1112/plms/pdt001.

[48] R. Eager, S. Franco, K. Schaeffer. Dimer Models and Integrable Systems. Journal of High Energy Physics, 06 (2012) 106.

[49] P. Edelman and C. Greene. Balanced tableaux. *Adv. in Math.* 63 (1987), no. 1, 42–99.

[50] N. Elkies. 1,2,3,5,11,37,...: Non-recursive solution found. Post to Sci.Math.Research, with an addendum by D. Speyer. http://faculty.uml.edu/jpropp/somos/elliptic.txt, 1995, accessed April 2012.

[51] A. Felikson, M. Shapiro and P. Tumarkin. Skew-symmetric cluster algebras of finite mutation type. J. Eur. Math. Soc. (JEMS) 14 (2012), no. 4, 1135–1180.

[52] A. Felikson, M. Shapiro and P. Tumarkin. Cluster algebras of finite mutation type via unfoldings. Int. Math. Res. Not. IMRN 2012, no. 8, 1768–1804.

[53] A. Felikson, M. Shapiro and P. Tumarkin. Cluster algebras and triangulated orbifolds. Adv. Math. 231 (2012), no. 5, 2953–3002.

[54] B. Feng, A. Hanany, and Y-H. He. D-brane gauge theories from toric singularities and toric duality. *Nucl. Phys. B*, 595 (2001), 165–200.

[55] V. Fock and A. Goncharov. Moduli spaces of local systems and higher Teichmüller theory. *Publ. Math. Inst. Hautes Études Sci.* No. 103 (2006), 1–211.

[56] V. V. Fock and A. B. Goncharov. Dual Teichmüller and lamination spaces, in *Handbook of Teichmüller Theory*. Vol. I, IRMA Lect. Math. Theor. Phys., 11, pp. 647–684. Eur. Math. Soc., Zürich, 2007.

[57] V. V. Fock and A. B. Goncharov. Cluster ensembles, quantization and the dilogarithm. *Ann. Sci. Éc. Norm. Supér.* (4) 42 (2009), no. 6, 865–930.

[58] V. V. Fock and A. B. Goncharov. The quantum dilogarithm and representations of quantum cluster varieties. *Invent. Math.*, 175(2):223–286, 2009.

[59] V.V. Fock and A.B. Goncharov. Cluster ensembles, quantization and the dilogarithm. II. The intertwiner. *In* Algebra, arithmetic, and geometry: in honor of Yu. I. Manin. Vol. I, volume 269 of Progr. Math., pages 655–673. Birkhäuser Boston Inc., Boston, MA, 2009.

[60] S. Fomin. Cluster algebras portal. http://www.math.lsa.umich.edu/~fomin/cluster.html, accessed June 2012.

[61] S. Fomin. Total positivity and cluster algebras. *Proceedings of the International Congress of Mathematicians*. Volume II, 125–145, Hindustan Book Agency, New Delhi, 2010.

[62] S. Fomin and P. Pylyavskyy, Tensor diagrams and cluster algebras. Preprint arxiv:1210.1888 [math.CO], 2012.

[63] S. Fomin and N. Reading. Generalized cluster complexes and Coxeter combinatorics. *Int. Math. Res. Not.* 2005, no. 44, 2709–2757.

[64] S. Fomin and N. Reading, Root systems and generalized associahedra. *Geometric combinatorics* 63–131, IAS/Park City Math. Ser., 13, Amer. Math. Soc., Providence, RI, 2007.

[65] S. Fomin, M. Shapiro, and D. Thurston. Cluster algebras and triangulated surfaces. I. Cluster complexes. *Acta Math.* 201 (2008), no. 1, 83–146.

[66] S. Fomin and D. Thurston. Cluster algebras and triangulated surfaces. II. Lambda lengths. Preprint arXiv:1210.5569, 2012.

[67] S. Fomin and A. Zelevinsky. Cluster algebras. I. Foundations. *J. Amer. Math. Soc.* 15 (2002), no. 2, 497–529 (electronic).

[68] S. Fomin and A. Zelevinsky. The Laurent phenomenon. *Adv. in Appl. Math.* 28 (2002), no. 2, 119–144.

[69] S. Fomin and A. Zelevinsky. Cluster algebras: notes for the CDM-03 conference. *Current developments in mathematics*, 2003, 1–34, Int. Press, Somerville, MA, 2003.

[70] S. Fomin and A. Zelevinsky. Cluster algebras. II. Finite type classification. *Invent. Math.* 154 (2003), no. 1, 63–121.

[71] S. Fomin and A. Zelevinsky. Y-systems and generalized associahedra. *Ann. of Math.* (2), 158 (2003), no. 3, 977–1018.

[72] S. Fomin and A. Zelevinsky. Cluster algebras. IV. Coefficients. *Compos. Math.* 143 (2007), no. 1, 112–164.

[73] A. P. Fordy, Mutation-periodic quivers, integrable maps and associated Poisson algebras. *Phil. Trans. R. Soc. A.* 369 (2011), no. 1939, 1264–1279.

[74] A. Fordy and A. Hone. Discrete integrable systems and Poisson algebras from cluster maps. Preprint arXiv:1207.6072, 2012.

[75] A. P. Fordy and R. J. Marsh. Cluster mutation-periodic quivers and associated Laurent sequences. *J. Algebraic Combin.* 34, no. 1 (2011), 19–66.

[76] W. Fulton. Young tableaux. With applications to representation theory and geometry. *London Mathematical Society Student Texts* 35. Cambridge University Press, Cambridge, 1997.

[77] D. Gale. The strange and surprising saga of the Somos sequences. *Math. Intelligencer* 13 (1991), 40–42.

[78] F. R. Gantmacher and M. G. Krein. Oscillation matrices and kernels and small vibrations of mechanical systems. AMS Chelsea Publishing, Providence, RI, 2002. (Original Russian edition, 1941.)

[79] C. Geiss, B. Leclerc and J. Schröer. Semicanonical bases and preprojective algebras. *Ann. Sci. Éc. Norm. Supér.* (4) 38 (2005), no. 2, 193–253.

[80] C. Geiß, B. Leclerc and J. Schröer. Kac-Moody groups and cluster algebras. *Adv. Math.* 228 (2011), no. 1, 329–433.

[81] C. Geiß, B. Leclerc and J. Schröer. Cluster structures on quantum coordinate rings. Selecta Mathematica (New Series) 19 (2013), 337–397.

[82] C. Geiß, B. Leclerc and J. Schröer. Factorial cluster algebras. Documenta Mathematica 18 (2013), 249–274.

[83] M. Gekhtman, M. Shapiro and A. Vainshtein. Cluster algebras and Poisson geometry. *Mathematical Surveys and Monographs*, 167. American Mathematical Society, Providence, RI, 2010.

[84] M. Gekhtman, M. Shapiro and A. Vainshtein. Cluster algebras and Weil-Petersson forms. *Duke Math. J.* 127 (2005), no. 2, 291–311.

[85] M. Gekhtman, M. Shapiro, and A. Vainshtein. On the properties of the exchange graph of a cluster algebra. Math. Res. Lett. 15 (2008), no. 2, 321–330.

[86] J. E. Grabowski and S. Launois. Quantum cluster algebra structures on quantum Grassmannians and their quantum Schubert cells: the finite-type cases. *Int. Math. Res. Not.* 2011, no. 10, 2230–2262.

[87] I. Grojnowski and G. Lusztig. A comparison of bases of quantized enveloping algebras. *In* Linear algebraic groups and their representations (Los Angeles, CA, 1992), volume 153 of Contemp. Math., pages 11–19. Amer. Math. Soc., Providence, RI, 1993.

[88] L. Guo. On tropical friezes associated with Dynkin diagrams. Int. Math. Res. Not. (2012), doi: 10.1093/imrn/rns176.

[89] R. Hartshorne, Algebraic geometry. *Graduate Texts in Mathematics* No. 52. Springer-Verlag, New York-Heidelberg, 1977.

[90] A. Henriques, A periodicity theorem for the octahedron recurrence. *J. Algebraic Combin.* 26 (2007), no. 1, 1–26.

[91] A. Henriques and J. Kamnitzer, The octahedron recurrence and gl_n crystals. *Adv. Math.* 206 (2006), no. 1, 211–249.

[92] C. Hohlweg. Permutahedra and Associahedra: Generalized associahedra from the geometry of finite reflection groups. *In* Associahedra, Tamari Lattices and Related Structures, Tamari Memorial Festschrift, editors F. Mueller-Hoissen, J. Pallo and J. Stasheff, Progress in Mathematics, vol. 299, (Birkhauser, 2012).

[93] A. N. W. Hone. Laurent polynomials and superintegrable maps. *SIGMA Symmetry Integrability Geom. Methods Appl.* 3, Paper 022, 18pp, 2007.

[94] Y. Hu and J. Ye. Canonical basis for type A4 (II) — polynomial elements in one variable. *Comm. Algebra* 33 (2005), no. 11, 3855–3877.

[95] Y. Hu, J. Ye and X. Yue, Canonical basis for type A4. I. Monomial elements. *J. Algebra* 263 (2003), no. 2, 228–245.

[96] The cluster complex of an hereditary Artin algebra. *Algebras and Representation Theory* 14 (2011), 14 (6), 1163–1185.

[97] J. E. Humphreys. Introduction to Lie algebras and representation theory. *Graduate Texts in Mathematics* Vol. 9. Springer-Verlag, New York-Berlin, 1972.

[98] J. E. Humphreys. Reflection groups and Coxeter groups. Volume 29 of *Cambridge Studies in Advanced Mathematics*. Cambridge University Press, Cambridge, 1990.

[99] R. Inoue and T. Nakanishi. Difference equations and cluster algebras I: Poisson bracket for integrable difference equations. Infinite analysis — 2010, Developments in quantum integrable systems, 63–88, RIMS Kôkyûroku Bessatsu, B28, Res. Inst. Math. Sci. (RIMS), Kyoto, 2011,

[100] N. Jacobson, Finite-dimensional division algebras over fields. Springer-Verlag, Berlin, 1996.

[101] M. Jimbo, A q-difference analogue of $U(g)$ and the Yang-Baxter equation. *Lett. Math. Phys.* 10:63–69, 1985.

[102] V. G. Kac. Infinite-dimensional Lie algebras. Third edition. Cambridge University Press, Cambridge, 1990.

[103] B. Keller. The periodicity conjecture for pairs of Dynkin diagrams. Ann. of Math. (2) 177 (2013), no. 1, 111–170.

[104] B. Keller. Categorification of acyclic cluster algebras: an introduction. Higher structures in geometry and physics, 227–241, *Progr. Math.* 287, Birkhäuser/Springer, New York, 2011.

[105] B. Keller. Cluster algebras, quiver representations and triangulated categories. Triangulated categories, 76–160, *London Math. Soc. Lecture Note Ser.* 375, Cambridge Univ. Press, Cambridge, 2010.

[106] B. Keller, Quiver mutation in Java. http://people.math.jussieu.fr/~keller/quivermutation/, 2006, accessed April 2012.

[107] B. Keller. On cluster theory and quantum dilogarithm identities. Representations of algebras and related topics, 85–116, EMS Ser. Congr. Rep., Eur. Math. Soc., Zürich, 2011.

[108] Y. Kimura. Quantum unipotent subgroup and dual canonical basis. *Kyoto J. Math.* 52 (2012), no. 2, 277–331.

[109] M. Kontsevich and Y. Soibelman. Motivic Donaldson-Thomas invariants: summary of results. Mirror symmetry and tropical geometry, 55–89, *Contemp. Math.* 527, Amer. Math. Soc., Providence, RI, 2010.

[110] M. Kontsevich and Y. Soibelman. Stability structures, motivic Donaldson-Thomas invariants and cluster transformations. Preprint arXiv:0811.2435v1 [math.AG], 2008.

[111] Y. Kodama and L. Williams. Combinatorics of KP solitons from the real Grassmannian. Preprint arXiv:1205.1101v1 [math.CO], 2012. To appear in the Proceedings of the Abel Symposium 2011, Balestrand, Norway.

[112] T. Lam and P. Pylyavskyy. Laurent phenomenon algebras. Preprint arXiv:1206.2611 [math.RT], 2012.

[113] T. Lam and P. Pylyavskyy. Linear Laurent phenomenon algebras. Preprint arXiv: 1206.2612v1 [math.RT], 2012.

[114] A. Lascoux and M.-P. Schützenberger. Structure de Hopf de l'anneau de cohomologie et de l'anneau de Grothendieck d'une variété de drapeaux. [Hopf algebra structures of the cohomology ring and Grothendieck ring of a flag manifold.] *C. R. Acad. Sci. Paris Sér. I Math.* 295 (1982), no. 11, 629–633.

[115] B. Leclerc. Cluster algebras and representation theory. *Proceedings of the International Congress of Mathematicians.* Volume IV, 2471–2488, Hindustan Book Agency, New Delhi, 2010.

[116] G. Lusztig. Canonical bases arising from quantized enveloping algebras. *J. Amer. Math. Soc.* 3 (1990), no. 2, 447–498.

[117] G. Lusztig, Introduction to Quantum Groups, Birkhuser, 1991.

[118] G. Lusztig. Quivers, perverse sheaves, and quantized enveloping algebras. *J. Amer. Math. Soc.* 4 (1991), no. 2, 365–421.

[119] G. Lusztig. Tight monomials for quantized enveloping algebras. *In* Quantum deformations of algebras and their representations, volume 7 of *Israel Math. Conf. Proc.* 117–132, 1993.

[120] G. Lusztig. Total positivity and canonical bases. Algebraic groups and Lie groups, 281–295, *Austral. Math. Soc. Lect. Ser.* 9, Cambridge Univ. Press, Cambridge, 1997.

[121] G. Lusztig. Semicanonical bases arising from enveloping algebras. *Adv. Math.* 151 (2000), no. 2, 129–139.

[122] J. L. Malouf. An integer sequence from a rational recursion. *Discrete Math.* 110 (1992), no. 1–3, 257–261.

[123] R. J. Marsh. More tight monomials for quantized enveloping algebras. *J. Algebra* 204 (1998), no. 2, 711–732.

[124] R. J. Marsh, M. Reineke and A. Zelevinsky. Generalized associahedra via quiver representations. *Trans. Amer. Math. Soc.* 355 (2003), no. 10, 4171–4186.

[125] R. J. Marsh and K. Rietsch. The B-model connection and mirror symmetry for Grassmannians. Preprint arXiv:1307.1085v1 [math.AG], 2013.

[126] R. J. Marsh and S. Schroll. A circular order on edge-coloured trees and RNA m-diagrams. Preprint arXiv:1010.3763v5 [math.CO], 2010. To appear in Advances in Applied Mathematics.

[127] H. Matsumoto. Générateurs et relations des groupes de Weyl généralisés. *C. R. Acad. Sci. Paris* 258 (1964), 3419–3422.

[128] E. Miller, and B. Sturmfels, Combinatorial commutative algebra. *Graduate Texts in Mathematics*, 227. Springer-Verlag, New York, 2005.

[129] S. Mukhopadhyay and K. Ray. Seiberg duality as derived equivalence for some quiver gauge theories. J. High Energy Phys. (2004) 02:070.

[130] G. Musiker. Perfect matchings and cluster algebras of classical type. 20th Annual International Conference on Formal Power Series and Algebraic Combinatorics (FPSAC 2008), 435–446, *Discrete Math. Theor. Comput. Sci. Proc. AJ* Assoc. Discrete Math. Theor. Comput. Sci., Nancy, 2008,

[131] G. Musiker and R. Schiffler. Cluster expansion formulas and perfect matchings. *J. Algebraic Combin.* 32 (2010), no. 2, 187–209.

[132] G. Musiker, R. Schiffler and L. Williams. Positivity for cluster algebras from surfaces. *Adv. Math.* 227 (2011), no. 6, 2241–2308.

[133] G. Musiker, R. Schiffler and L. Williams. Bases for cluster algebras from surfaces. Compos. Math. 149 (2013), no. 2, 217–263.

[134] G. Musiker and C. Stump, A compendium on the cluster algebra and quiver package in Sage. *Sém. Lothar. Combin.* 65 (2010/12), Art. B65d, 67 pp.

[135] K. Nagao. Donaldson-Thomas theory and cluster algebras. Duke Math. J. 162 (2013), no. 7, 1313–1367.

[136] T. Nakanishi. Periodicities in cluster algebras and dilogarithm identities. *In* Representations of algebras and related topics (A. Skowronski and K. Yamagata, eds.), EMS Series of Congress Reports, European Mathematical Society, 2011, pp.407–444,

[137] T. Nakanishi. Tropicalization method in cluster algebras. Tropical geometry and integrable systems, 95–115, Contemp. Math., 580, Amer. Math. Soc., Providence, RI, 2012.

[138] K. Nagao, Y. Terashima and M. Yamazaki. Hyperbolic 3-manifolds and cluster algebras. Preprint arXiv:1112.3106 [math.GT], 2011.

[139] S. Oh, A. Postnikov and D. E. Speyer. Weak separation and plabic graphs. Preprint arXiv:1109.4434v1 [math.CO], 2011.

[140] The On-Line Encyclopedia of Integer Sequences, published electronically at http://oeis.org, 2012.

[141] D. I. Panyushev. ad-nilpotent ideals of a Borel subalgebra: generators and duality. *J. Algebra* 274 (2004), no. 2, 822–846.

[142] P.-G. Plamondon. Generic bases for cluster algebras from the cluster category. Int. Math. Res. Notices (2012) doi:10.1093/imrn/rns102.

[143] A. Postnikov. Total positivity, Grassmannians, and networks. Preprint arXiv:math/0609764v1 [math.CO], 2006.

[144] J. Propp. The combinatorics of frieze patterns and Markoff numbers. Preprint arXiv:math/0511633v4 [math.CO], 2005.

[145] J. Propp, A bare-bones chronology of Somos sequences. http://faculty.uml.edu/jpropp/somos.html, July 2006, accessed April 2012.

[146] I. Reiten. Cluster categories. *Proceedings of the International Congress of Mathematicians.* Volume I, 558–594, Hindustan Book Agency, New Delhi, 2010,

[147] I. Saleh. On the automorphisms of cluster algebras. Preprint arXiv:1011.0894v1 [math.RT], 2010.

[148] H. Samelson. Notes on Lie algebras. Second edition. *Universitext*. Springer-Verlag, New York, 1990.

[149] R. Schiffler. A cluster expansion formula (A_n case). *Electron. J. Combin.* 15 (2008), no. 1, Research paper 64, 9 pp.

[150] R. A. Schiffler. A geometric model for cluster categories of type D_n. *J. Algebraic Combin.* 27 (2008), no. 1, 1–21.

[151] R. Schiffler. On cluster algebras arising from unpunctured surfaces. II. *Adv. Math.* 223 (2010), no. 6, 1885–1923.

[152] R. Schiffler and H. Thomas. On cluster algebras arising from unpunctured surfaces. *Int. Math. Res. Not.* 2009, no. 17, 3160–3189.

[153] J. S. Scott, Grassmannians and cluster algebras. *Proc. London Math. Soc.* (3) 92 (2006), no. 2, 345–380.

[154] A. I. Seven. Recognizing cluster algebras of finite type. *Electron. J. Combin.* 14 (2007), no. 1, Research Paper 3, 35pp. (electronic).

[155] D. E. Speyer, Perfect matchings and the octahedron recurrence. *J. Algebraic Combin.* 25 (2007), no. 3, 309–348.

[156] R. P. Stanley. On the number of reduced decompositions of elements of Coxeter groups. *European J. Combin.* 5 (1984), no. 4, 359–372.

[157] R. P. Stanley. Enumerative combinatorics. Vol. 2. With a foreword by Gian-Carlo Rota and Appendix 1 by Sergey Fomin. Cambridge Studies in Advanced Mathematics, 62. Cambridge University Press, Cambridge, 1999.

[158] J. D. Stasheff, Homotopy associativity of H-spaces. I, II. *Trans. Amer. Math. Soc.* 108 (1963), 275–292; ibid. 108 (1963) 293–312.

[159] B. Sturmfels. Gröbner bases and convex polytopes. Volume 8 of *University Lecture Series*. American Mathematical Society, Providence, RI, 1996.

[160] J. Tits. Le problème des mots dans les groupes de Coxeter. 1969 *Symposia Mathematica* (INDAM, Rome, 1967/68), Vol. 1, pp. 175–185, Academic Press, London.

[161] J. Vitória. Mutations vs. Seiberg duality. *J. Algebra* 321 (2009), no. 3, 816–828.

[162] N. Xi. Canonical basis for type A_3. *Comm. Algebra* 27 (1999), no. 11, 5703–5710.

[163] N. Xi. Canonical basis for type B_2. *J. Algebra* 214 (1999), no. 1, 8–21.

[164] Dong Yang, Clusters in non-simply-laced finite type via Frobenius morphisms. Algebra Colloq. 16 (2009), no. 1, 143–154.

[165] Al. B. Zamolodchikov. On the thermodynamic Bethe ansatz equations for reflectionless ADE scattering theories. *Phys. Lett. B* 253 (1991), no. 3–4, 391–394.

[166] B. Zhu. BGP-reflection functors and cluster combinatorics. *J. Pure Appl. Algebra* 209 (2007), no. 2, 497–506.

[167] B. Zhu. Preprojective cluster variables of acyclic cluster algebras. *Comm. Algebra* 35 (2007), no. 9, 2857–2871.

[168] A. Zelevinsky. From Littlewood-Richardson coefficients to cluster algebras in three lectures. Symmetric functions 2001: surveys of developments and perspectives, 253–273, NATO Sci. Ser. II Math. Phys. Chem., 74, Kluwer Acad. Publ., Dordrecht, 2002.

[169] A. Zelevinsky. Cluster algebras: origins, results and conjectures. *Advances in algebra towards millennium problems*, 85–105, SAS Int. Publ., Delhi, 2005.

[170] A. Zelevinsky. What is ... a cluster algebra? *Notices Amer. Math. Soc.* 54 (2007), no. 11, 1494–1495.

[171] G. M. Ziegler. Lectures on polytopes. Volume 152 of *Graduate Texts in Mathematics*. Springer-Verlag, New York, 1995.

Nomenclature

$(-,-)$	bilinear form, page 31
$(-\|-)_Q$	Q-compatibility degree, page 58
α	root, page 33
α_i	simple root, page 36
Δ	simple system, page 35
Γ	Dynkin diagram or corresponding valued graph, page 39
κ_F	normal cone of a face F of a polytope, page 70
λ	partition, page 45
μ_k	mutation at k, page 11
Φ, Φ_W	root system, page 33
Φ^+	positive system, page 35
$\pi_{k,n}$	permutation sending i to $i + k \mod n$, page 95
ρ	the permutation $(1\ 2 \cdots n)$, page 76
σ_i	truncated simple reflection, page 58
$[a, b]$	$\{a, a+1, \dots, b\}$, page 9
$\mathcal{A}(\widetilde{\mathbf{x}}, \widetilde{B})$	cluster algebra with initial seed $(\widetilde{\mathbf{x}}, \widetilde{B})$, page 11
$A = (a_{ij})$	Cartan matrix, page 37
$\mathcal{A}(D)$	cluster algebra associated to a Postnikov diagram D, page 96
$\widetilde{B} = (b_{ij})$	exchange matrix, page 10
$B = (b_{ij})$	principal part of exchange matrix, page 10
$B(t)$	matrix in exchange pattern, page 23
$\overline{B}$	folded exchange matrix, page 53
C	Q-root cluster, page 58
$\mathbf{c}$	coefficients, page 10
c_Δ	Coxeter element, page 47
$Cl(B)$	family of cluster algebras, page 49
$C(t)$	non-principal part of exchange matrix in geometric exchange pattern, page 27
$\mathbb{C}[Y]$	homogeneous coordinate ring of Y, page 92
D	Postnikov diagram, page 94
d_i	degree of reflection group, page 48
(E1)-(E4)	axioms for exchange pattern, page 20

e_i	exponent of reflection group, page 48
$\mathbb{F}$	field of rational functions, page 10
$Gr(k,n)$	Grassmannian, page 91
H_α	hyperplane orthogonal to α, page 32
I	the set $\{1,2,\dots,n\}$, page 19
K	a k-subset of I, page 95
$m(i,j)$	number in Coxeter presentation, page 36
$M_j(t)$	monomial in exchange pattern, page 19
$\mathbb{P}$	coefficient group, page 19
$\mathbf{p}(D)$	Plücker coordinates associated to a Postnikov diagram D, page 96
$p_{i_1,i_2,\dots,i_k}$	Plücker coordinate, page 91
$p_j(t)$	coefficient in exchange pattern, page 19
P_n	regular polygon with n vertices, page 93
$P_n^{(t)}$	tth primitive quiver with n vertices, page 76
Q	quiver, page 15
Q	valued quiver, page 16
Q_{alt}	an alternating orientation of Γ, page 65
$Q(D)$	quiver of Postnikov diagram D, page 96
$(Q,\mathbf{Y})$	Y-seed, page 82
$\widetilde{Q}$	quiver with frozen vertices, page 14
$\overline{Q}$	folded (valued) quiver, page 54
s_α	reflection in H_α, page 32
$S(\gamma)$	support of a root, page 66
$S_+(\gamma)$	positive support of a root, page 66
$S_-(\gamma)$	negative support of a root, page 66
$\operatorname{sgn}(x)$	sign of x, page 9
s_i	simple reflection, page 36
(S,M)	marked Riemann surface, page 84
$S(X)$	symmetric algebra of a vector space X, page 48
$\mathcal{T}$	tagged triangulation of a surface, page 87
$\mathcal{T}$	triangulation of a surface, page 85
$\mathbb{T}_n$	n-regular tree, page 19
$T(X)$	tensor algebra of a vector space X, page 48
V	Euclidean space, page 31

W	finite reflection group, page 33
w_0	longest element of a finite reflection group, page 45
$[x]_+$	$\max(x, 0)$, page 9
$[x]_-$	$\min(0, x)$, page 9
$x_i(t)$	cluster variable in exchange pattern, page 19
$\mathbf{x}(t)$	cluster in exchange pattern, page 19
$\mathbf{x}$	cluster, page 10
$\widetilde{\mathbf{x}}$	extended cluster, page 10
$(\widetilde{\mathbf{x}}, \widetilde{B})$	seed, page 10
$Y_{i,i',t}$	Y-system, page 81

Index

$\pi-$ diagram, 94
admissible group action, 53
algebra
    exterior, 90
    symmetric, 48
    tensor, 48
alternating strand diagram, 94
arc, 85
    tagged, 87
        compatibility, 87
Artin group, 46
associahedron, 72
    generalized, 70
    type A_n, 71

bilinear form, 31
    positive definite, 31
    symmetric, 31
braid group, 46
braid relation, 46

Cartan counterpart, 50
Cartan matrix, 37
    finite type, 37
categorical periodicity, 81
cluster, 10
    extended, 10
    number, 62
    positive, 62
    root, 58
cluster algebra, 10, 11
    acyclic, 57
    classification, 50, 51
    exchange pattern, 22
    family, 49
    finite mutation type, 49
    finite type, 18, 49, 50
    geometric type, 12, 30
    Ptolemy, 94
    quiver notation for, 13
    strong isomorphism, 17, 49
    type, 52
    type A_3, 56
cluster category, 81
    generalized, 81
cluster complex, 68
    sphere, 68
cluster expansion, 66
coefficient, 10
    principal, 10
coefficient group, 19
commutation class, 47
    number, 47
commutation relation, 47
compatibility degree, 58
cone, 63
    affine, 92
    face, 63
    polyhedral, 63
    proper face, 63
    simplicial, 64
convex hull, 67
Coxeter element, 47
Coxeter graph, 43, 44
Coxeter group, 36
    irreducible, 43
Coxeter number, 47
Coxeter presentation, 36, 46
cyclohedron, 72

degree, 48
del Pezzo surface, 80
Denominator Theorem, 55
dihedral group, 32
Dynkin diagram, 39, 40
    simply laced, 51

Euclidean space, 31
exchange graph, 17
exchange matrix, 10
exchange pattern, 19
    geometric type, 27

exchange polynomial, 19
exchange relation, 11, 19
exponent, 47
exterior power, 90

fan, 63
 cluster, 65
 complete, 64
 simplicial, 64
flip, 72, 88
folding, 53
frozen variable, 10
frozen vertex, 14

Gale-Robinson recurrence, 78, 79
geometric exchange, 97
Grassmannian, 91
 $Gr(2, n)$, 92, 98
 finite cluster type, 98

homogeneous coordinate ring, 92
hook, 45
 length, 45
hyperbolic geometry, 84

Laurent phenomenon, 76
Laurent sequence, 76
longest element, 45

matrix
 sign-skew-symmetric, 10
 skew-symmetric, 13
 skew-symmetrizable, 13
 symmetrizable, 37

octahedron recurrence, 76
orthogonal transformation, 31

partition, 45
pentagon recurrence, 8
Plücker coordinate, 91
Plücker embedding, 92
Plücker relation, 91
polytope, 69
 face, 69
  normal cone, 69
 normal fan, 70
 proper face, 69
 simple, 70
 Stasheff, 72
positive system, 35
Postnikov arrangement, 94
Postnikov diagram, 94
principal part, 10
Ptolemy's Theorem, 94
puncture, 84

Q-compatible, 59
quiver
 finite mutation type, 84
  classification, 84
 from Riemann surface, 84
 period 1, 76, 79
  classification, 77
 period a, 80
 primitive, 77
 tensor product, 81
 triangle tensor product, 81
 valued, 16, 51
quiver gauge theory, 80
quiver mutation, 14

reduced expression, 36, 44
 length, 36
 number, 47
reflection, 31
 formula for, 32
 simple, 35
 truncated simple, 58
reflection functor, 59
reflection group, 32
 crystallographic, 37
 irreducible, 33
 type A_n, 39
region
 alternating, 95
 oriented, 95
root
 negative, 35
 positive, 35
 simple, 35

root cluster, 57
negative support, 64
positive, 62
root system, 33
classification, 39
crystallographic, 37
irreducible, 33
isomorphism, 34
type A_2, 41
type A_n, 39
type B_2, 42
type G_2, 42

seed, 10
equivalence, 10
mutation, 11
of quiver type, 15
of valued quiver type, 16
similarity, 31
simple system, 35
simplex, 67
simplicial complex
abstract, 68
realization, 68
Euclidean, 67
sinks
adapted sequence, 59
admissible sequence, 59
Somos sequence, 8, 76, 79
stable variable, 10
strongly isomorphic, 49
symmetric group, 46

Teichmüller space, 84
tree
regular, 19
triangle
self-folded, 86
triangulation
ideal, 85
reachable, 88
tagged, 87
zig-zag, 71

valued graph, 16
vector
decomposable, 91
length, 31

Y-pattern, 82
restricted, 82
Y-system, 81
Young diagram, 45

Zamolodchikov periodicity conjecture, 81